ISBN 978-1-334-03375-9
PIBN 10595073

1 MONTH OF
FREE
READING

at

www.ForgottenBooks.com

By purchasing this book you are eligible for one month membership to ForgottenBooks.com, giving you unlimited access to our entire collection of over 700,000 titles via our web site and mobile apps.

To claim your free month visit:

www.forgottenbooks.com/free595073

English
Français
Deutsche
Italiano
Español
Português

www.forgottenbooks.com

Mythology Photography **Fiction**
Fishing Christianity **Art** Cooking
Essays Buddhism Freemasonry
Medicine **Biology** Music **Ancient
Egypt** Evolution Carpentry Physics
Dance Geology **Mathematics** Fitness
Shakespeare **Folklore** Yoga Marketing
Confidence Immortality Biographies
Poetry **Psychology** Witchcraft
Electronics Chemistry History **Law**
Accounting **Philosophy** Anthropology
Alchemy Drama Quantum Mechanics
Atheism Sexual Health **Ancient History**
Entrepreneurship Languages Sport
Paleontology Needlework Islam
Metaphysics Investment Archaeology
Parenting Statistics Criminology
Motivational

The Zodiacal Light.—p. 178.

NATURAL PHENOMENA

A COLLECTION OF

𝔇escriptive and 𝔖peculative 𝔈ssays

ON SOME OF THE

𝔅y=paths of 𝔑ature

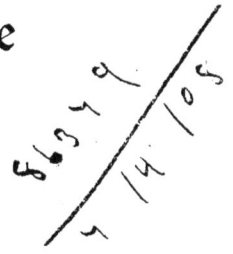

BY

F. A. BLACK

Author of "Terrestrial Magnetism and its Causes."

𝔏ondon:

GALL AND INGLIS, 25 PATERNOSTER SQUARE: E.C.

AND EDINBURGH.

They who are accustomed to studies of this nature will expect, and they will allow, too, for many faults. They know that many of the objects of our enquiry are in themselves obscure and intricate . . . The characters of Nature are legible, it is true, but they are not plain enough to enable those who run to read them. We must make use of a cautious, I had almost said a timorous, method of proceeding . . . If an enquiry thus carefully conducted, should fail at last of discovering the truth, it may answer an end perhaps as useful in discovering to us the weakness of our own understanding . . . If it does not preserve us from error it may at least from the spirit of error, and may make us cautious of pronouncing with positiveness or with haste, when so much labour may end in so much uncertainty.

EDMUND BURKE,
On the Sublime and Beautiful.

Few of His works can we survey :
These few our skill transcend :
But the full thunder of His power
What heart can comprehend?

Job. xxvi. 14, (paraphrased).

Preface.

——o——

The following Essays, though they all relate to what may reasonably be described as "Natural Phenomena,"—or the Methods in which the Works, or Ways, of Nature are made manifest,—may conveniently, according to subject matter, be divided into three classes. There are four Essays the subjects of which may be described as meteorological; three, which deal more or less with celestial phenomena, or at least with the Earth only in association with other members of the Solar System; and three, the subjects of which are not only exclusively of a terrestrial, but practically of a geographical, character.

The sequence adopted gives an Essay of each class in turn, the Essays of the same class not being grouped together. Thus, the first Essay is distinctly of a meteorological type, the second is geographical, and the third is concerned with other members of our system as well as with the Earth.

The subjects treated may be considered as being, to some extent, disconnected. Yet, in addition to their mutual relation to Nature, they have a connecting link of an interesting sort. It may be said that each subject has a certain mystery associated with it, a mystery which, in some cases, is as puzzling to scientific men as to the mere "man in the street."

I have endeavoured throughout to be as clear as possible and to avoid all technicalities. In any instance, where, in dealing with matters in regard to which scientific opinion is uncertain, I have ventured to suggest an hypothesis of any

PREFACE.

novelty, I have invariably given the whole facts which appear to bear upon the subject, in so far as these are requisite to enable the reader to form an independent judgment.

I have to acknowledge my indebtedness to Mr. Arthur H. MacKenzie, M.A., B.Sc., of the Central High School, Leeds, for kindly going most carefully over the manuscript, and making various suggestions of much value.

I have also to thank the Publishers for much assistance.

FRED^{K.} A. BLACK.

59 ACADEMY STREET,
INVERNESS. 13TH JULY, 1906.

Contents.

———o———

CONTENTS.

VIII.—*THE DAY AND THE PLACE OF ITS BIRTH.*

IX.—*THE ROTATION OF THE EARTH AND OTHER
MEMBERS OF THE SOLAR SYSTEM.*

The stupendous importance of the Earth's rotation
and our ignorance as to its causes—Period of Earth's
rotation and uncertainty as to its absolute constancy—

X.—*THE DAILY BAROMTERIC TIDE.*

LIST OF DIAGRAMS AND ILLUSTRATIONS.

——o——

LIST OF DIAGRAMS AND ILLUSTRATIONS.

LIST OF TABLES.

——o——

THE TURN OF THE DAY.

SYNOPSIS.

Definition and appropriateness of the phrase—Analogy between the daily and the yearly changes in the power of the Sun—Maximum solar influence, whether diurnal or annual, not coincident with maximum temperature of the air—Daily and annual temperature oscillations considered—Distinction between the northern and southern hemispheres in the annual temperature oscillation—Modes of transference of heat—Bodies permeable to heat—Relative absorption of heat by the atmosphere, the sea, and the land—Heating of the atmosphere by terrestrial convection—Direct and indirect solar influence—Variation of temperature of atmosphere at different levels—Variation in time of occurrence of daily maximum temperature according to position and season—Ambiguity of the terms "Summer" and "Winter"—The bearing of the subject on the science of meteorology.

THE TURN OF THE DAY.

"THE Turn of the Day" is a colloquialism which is singularly appropriate in its application. It has reference to the periods, which, with more profundity, if not greater accuracy, we describe as "the solstices." These are the times when, after a long spell of gradual increase in the duration of daylight, the day "turns" in the opposite direction, and the period of daylight begins to shorten; and when, after a long spell of gradual increase in the duration of darkness, the day "turns," the night begins to shorten, and the daily visits of the Sun become more prolonged. On these occasions, which occur about the 21st June and the 22nd of December annually, the Sun ceases its northward progress, and "turns" southward again, or ceases its southward progress and again "turns" northward. In different aspects, therefore, the phrase is apt, as relating to what may reasonably be described as a "turning" movement.

In a more limited sense, we may fittingly use the expression, "the turn of the day," with reference to the *daily* occurrence of solar changes. The Sun daily rises above the eastern horizon, crosses the meridian, and sinks below the western horizon. The daily crossing of the meridian by the Sun, marks, in this sense, "the turn of the day."

There are many respects, indeed, in which, outside the tropics, the solar changes relating to the day have a close analogy to the solar changes relating to the year. Daily, as well as yearly, we have the Sun mounting upwards in the heavens, reaching its greatest elevation, and slowly descending. We may compare the crossing of the meridian by the Sun, midway in its daily visible journey, to its reaching its highest point in the heavens at the summer solstice, midway between its crossing of the equator in its progress into our hemisphere, and its subsequent crossing of the equator on its return to the opposite hemisphere. In the day, as well as in the year, the power of the Sun gradually increases, reaches its maximum, and gradually declines.

Now, as the Sun has undoubtedly the predominating influence in the production of terrestrial heat, in so far as dwellers on the surface of the Earth are concerned, it might naturally be concluded that, when the power of the Sun reaches its maximum, the temperature of the atmosphere should also reach its maximum. As the Sun, in its daily journey, attains its greatest influence at noon, when it is crossing the meridian, we should expect the temperature to reach its highest level for the day at the hour of noon. As the Sun, in its *annual* journey, attains its greatest influence at the summer solstice, when—at least outside the tropics—it is nearest to the zenith, we should expect the temperature to reach its highest level for the year about the 21st of June

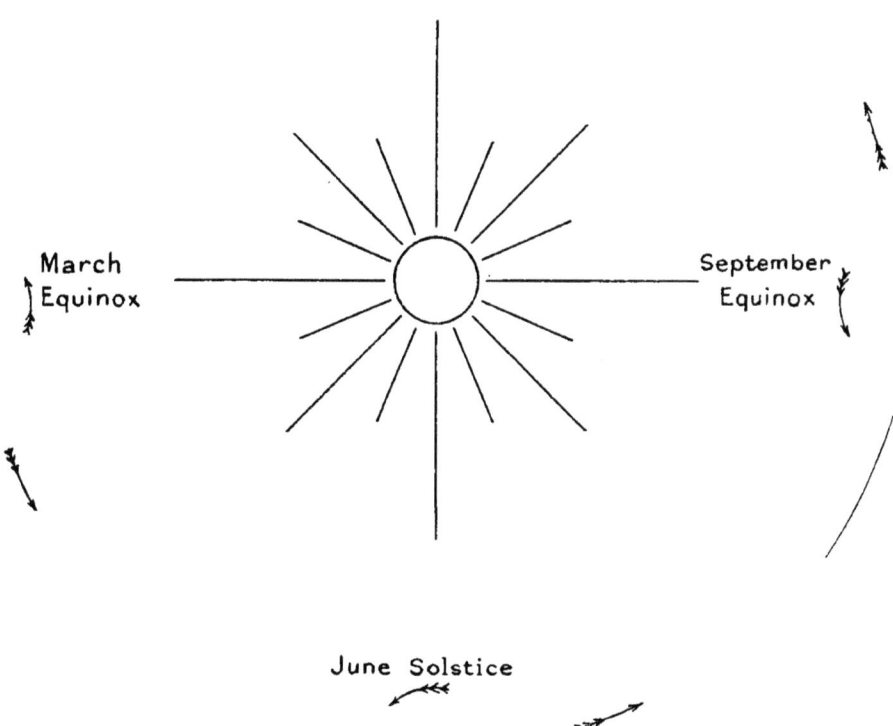

Diagram showing "The Turn of the Day" in its diurnal and annual aspects.—p. 4.

in the Northern Hemisphere, and about the 22nd of December in the Southern Hemisphere.

There is, certainly, no doubt that the Sun is the main source of terrestrial heat. With increase of exposure to the Sun, we have, generally speaking, increase of temperature, and with lessened exposure we have fall of temperature. It is also evident that the Sun—if its declination and the state of the atmosphere are similar—has greater influence when on the meridian than when it is in any other position; and that, under the same atmospheric conditions, its influence is greatest when its position on the meridian is at the zenith, and becomes less according to the distance of separation from the zenith. The late Mr. R. A. Proctor expresses the matter in the following passage :—

"As the Sun is perceptibly the source of light and heat, it follows that when he is north of the equator, we receive (in our northern latitudes) more light and heat than when he is on the equator, and so much the more as his northerly declination is greater; while when he is south of the equator, we receive less light and heat than when he is on the equator, and so much the less as his southerly declination is greater."

We know that the difference in the seasons is not occasioned by the greater or less distance of the Earth from the Sun, although this has an effect in modifying or intensifying the character of the seasons. The difference is occasioned by the more or less oblique direction of the Sun's rays. In the summer of the Northern Hemisphere, the Earth is much more distant from the Sun than it is in the winter of that hemisphere, and, consequently, neither the heat of summer nor the cold of winter is so extreme in the Northern, as it is in the Southern Hemisphere. When the Earth is at nearly its greatest distance from the Sun, its rays fall most directly upon us in these northern latitudes, and therefore we have summer, while the obliqueness of the rays when the Earth is nearest to the Sun gives us winter.

In the *daily* course of the Sun, its rays are certainly most direct when it is crossing the meridian, just as in its *annual* course they are most direct at the summer solstice. In the daily course, the obliquity of the rays increases with separation from the meridian, just as in the annual course, the obliquity increases with separation from the summer solstice. Thus there is apparently strong reason to suppose that the daily temperature should reach its maximum at the hour of noon, and that the annual temperature should reach its maximum at the summer solstice.

Yet observation goes to show the correctness of the popular belief that the time of occurrence of the maximum solar influence, does not generally coincide with the time of occurrence of the maximum temperature. In the daily march of temperature, it would appear that the highest point is not, as a general rule, attained until an appreciable interval after the Sun has crossed the meridian; and, in the annual march of temperature, the highest point occurs some time after the summer solstice, and the lowest point some time after the winter solstice. Meteorological observation would seem to prove the accuracy of the common supposition that the warmest weather, and the coldest weather of the year, do not coincide with, but that, in each case, they follow "the

turn of the day." In this respect, as in other respects
the daily phenomena would seem to bear a close
resemblence to the annual phenomena.

Let us now consider the observations which have
been made in regard to the daily and annual
temperature oscillations, dealing with them in the order
mentioned.

TABLE SHOWING HOURLY DEVIATIONS FROM MEAN DAILY
TEMPERATURE AT ROTHESAY AND BATAVIA.

Hour.	Rothesay.	Batavia.	Hour.	Rothesay.	Batavia.
1 A.M.	-- 1·7° F.	− 3·2° F.	1 P.M.	+ 2·4° F.	+ 5·7° F.
2 ,,	− 2·0°	− 3·6°	2 ,,	+ 2·7°	+ 5·6°
3 ,,	− 2·1°	− 4·0°	3 ,,	+ 2·8°	+ 5·2°
4 ,,	− 2·2°	− 4·3°	4 ,,	+ 2·6°	+ 4·3°
5 ,,	− 2·2°	− 4·7°	5 ,,	+ 2·1°	+ 3·3°
6 ,,	− 2·0°	− 4·9°	6 ,,	+ 1·5°	+ 1·9°
7 ,,	− 1·5°	− 4·3°	7 ,,	+ 0·9°	+ 0·6°
8 ,,	− 0·9°	− 2·2°	8 ,,	+ 0·2°	+ 0·4°
9 ,,	− 0·2°	− 0·5°	9 ,,	− 0·4°	− 1·2°
10 ,,	+ 0·5°	+ 2·8°	10 ,,	− 0·8°	− 1·8°
11 ,,	+ 1·2°	+ 4·4°	11 ,,	− 1·2°	− 2·3°
Noon	+ 1·9°	+ 5·4°	M'dn'gt	− 1·5°	− 2·8°

Dr. Alexander Buchan in his treatise on "Meteorology,"
in the current edition of the *Encyclopædia Britannica,*
gives a table showing the hourly deviations from the
mean daily temperature, at two stations which may be
accepted as being fairly typical. These stations are
Rothesay in the Island of Bute, on our own western
coast, and Batavia, in the Dutch East Indies. The
former is in the Northern Hemisphere, in a high temperate
latitude : the latter is in the Southern Hemisphere, in a

low tropical latitude. Both, however, are near the sea. Rothesay is in latitude 55° 50′ N., and longitude 5° 4′ W., and its mean annual temperature is 47·3° F. Batavia is in latitude 6° 8′ S., and longitude 106° 48′ E., and its mean annual temperature is 78·7° F.

TABLE SHOWING HOUR OF OCCURRENCE OF MAXIMUM DAILY TEMPERATURE AT CERTAIN STATIONS IN THE NORTHERN HEMISPHERE, IN SUMMER AND WINTER RESPECTIVELY.

Town.	Geographical Situation.	Hours of occurrence of Maximum Temperature.	
		Summer.	Winter.
Archangel	64° 32′ 8″ N., 40° 33′ E.	2.50 P.M.	1.30 P.M.
Bogoslovsk	*Abt.* 59° 45′ N., 60° 10′ E.	2.5 ,,	1.30 ,,
Sitka	,, 57° N., 135° 20′ W.	0.50 ,,	1.30 ,,
Rothesay	55° 50′ N., 5° 4′ W.	3.15 ,,	2.30 ,,
Oxford	51° 45′ N., 1° 15′ W.	3.10 ,,	2.0 ,,
Geneva	*Abt.* 46° 12′ N., 6° 10′ E.	2.50 ,,	2.0 ,,
St. Bernard	45° 51′ N., 7° 11′ 23″ E.	1.20 ,,	0.55 ,,
Toronto	43° 39′ 35″ N., 79° 23′ 39″ W.	3.45 ,,	1.50 ,,
Tiflis	41° 42′ N., 44° 48′ E.	3.10 ,,	2.25 ,,
Madrid	40° 24′ 35″ N., 3° 41′ 51″ W.	2.50 ,,	2.40 ,,
Philadelphia	39° 57′ 7″ N., 75° 9′ 23″ W.	3.10 ,,	2.40 ,,
Calcutta	22° 33′ 47″ N., 88° 23′ 34″ E.	0.40 ,,	2.30 ,,
Bombay	18° 53′ 54″ N., 72° 52′ E.	1.30 ,,	2.10 ,,
Madras	13° 4′ 8″ N., 80° 14′ 51″ E.	1·25 ,,	0.40 ,,

Dr. Buchan mentions that in Batavia the maximum temperature was found to occur at 1.20 p.m., the same phase taking place in Rothesay at about 3 p.m. At Batavia, where the length of the day does not vary to any great extent, the hour of the maximum is fairly constant. In Rothesay, on the other hand, where the length of the day varies greatly in different seasons, the time of occurrence of the maximum has considerable variation.

The table on page 8 shows the hours at which the highest temperature for the day occurs, in summer and winter respectively, at various stations in the Northern Hemisphere, the stations being arranged in order of latitude, commencing farthest north.

It will be noticed that at every station without exception, the maximum temperature for the day occurs a considerable time after the Sun has passed the meridian, the smallest interval being 40 minutes. This is the interval at Calcutta in summer, and at Madras in winter. It would appear that the length of time which elapses between the crossing of the meridian by the Sun and the occurrence of the maximum temperature for the day, has no noticeable dependence on latitude except as regards seasonal variation.

There are stations in the higher latitudes, as well as in the lower latitudes, between which there is great diversity in the length of the time which intervenes. The mean time of the occurrence of the maximum temperature at these fourteen stations is about 2.21 p.m. in summer, and 1.55 p.m. in winter. It would seem that the longer duration of sunshine in the summer days has a decided influence in postponing the hour of maximum temperature, as this postponement is most evident in the higher latitudes, being the latitudes where the increased length of the day in summer as compared with winter, is most obvious.

These tables apply exclusively to the land surface. Some information in regard to the daily march of temperature over the open sea was obtained by the

Challenger Expedition.   The above table records the mean deviation of the temperature of the air in each succeeding two hours over the North Atlantic Ocean.   It is compiled from observations made by the *Challenger* on 126 days, partly from March to August 1873, and partly in April and May 1876, the mean position at which the observations were made being about 30° N. latitude, and 42° W. longitude.

TABLE SHOWING BI-HOURLY DEVIATIONS FROM MEAN DAILY TEMPERATURE OVER THE NORTH ATLANTIC OCEAN, AS ASCERTAINED BY THE *CHALLENGER* EXPEDITION.

Hour.	Deviation from Daily Mean.	Hour.	Deviation from Daily Mean.
2 A.M.	− 1·13°	2 P.M.	+ 1·80°
4 ,,	− 1·40°	4 ,,	+ 1·56°
6 ,,	− 1·41°	6 ,,	+ 0·73°
8 ,,	− 0·21°	8 ,,	− 0·30°
10 ,,	+ 0·78°	10 ,,	− 0·80°
Noon	+ 1·45°	Midnight	− 1·02°

This table goes to show that, as regards the occurrence of a considerable interval of time between the crossing of the meridian by the Sun and the attainment of the maximum temperature for the day, the sea surface and the land surface fall into the same category.   In both, the interval occurs, and, in the ocean records, the time of occurrence of the maximum temperature comes out at an hour corresponding to the mean time of its occurrence on the land surface, as revealed by the observations made at the stations mentioned in the preceding table. The mean time of the occurrence of the maximum at these stations, taking summer and winter together, is about two hours eight minutes after the Sun has passed its highest point for the day; while the *Challenger* records show the corresponding time in the North Atlantic Ocean to be about two hours. Thus in both cases the maximum temperature appears on the average to occur daily about 2 p.m.

It is, however, not invariably the case that an appreciable interval occurs between the time when the Sun attains its greatest power for the day and the time when the thermometer reaches its highest point in its daily oscillation. Dr. Buchan mentions that, about the time when the observations were made in the North Atlantic upon which the table of the bi-hourly deviations is based, the *Challenger* was for 76 days lying near the land. Observations were also taken on these days, which, therefore, relate to the daily temperature movement on the margin of a great ocean. Dr. Buchan continues :—

"The observations made on these days show a greater daily range of temperature of the air than occurred out in the open sea. The minimum – 2·05°, occurred at 4 a.m., and the maximum, + 2·33°, at noon, thus giving a daily range of 4·38°. The occurrence of the maximum so early as noon is doubtless occasioned by the greater strength of the sea breeze after this hour, this maintaining a lower temperature."

Thus the coincidence of the maximum daily power of the Sun and the maximum daily temperature is quite exceptional, and is explained by local causes. In general the maximum temperature for the day occurs a considerable time—being on the average about two hours—after the Sun has passed the meridian; and, in this respect, there appears to be no difference between the Northern and Southern Hemispheres.

In the *annual*, as distinguished from the *daily*, temperature oscillation, a material difference arises between the two hemispheres. Not only are the seasons reversed, but the continuance of the Sun is not the same in each hemisphere. As the Earth moves in its orbit with less velocity as its distance from the Sun increases, the interval between the March equinox and the September equinox—which is the time during which the Sun is north of the equator—is greater than the interval between the September equinox and the March equinox—which is the time during which the Sun is south of the equator. We may take the date of the vernal and autumnal equinoxes of the Northern Hemisphere to be the 21st of March and the 23rd of September, respectively. The period between these dates is 186 days, while from 23rd September to 21st March following is only 179 days. Thus the Sun each year is north of the equator something like seven days longer than it is south of the equator.

The diversity in the heating of the two hemispheres which must result from (1) the Northern Hemisphere having its winter when the Earth is nearest to the Sun, and its summer when the earth is farthest away from the Sun, while the converse is the case in the Southern Hemisphere, and (2) the Northern Hemisphere being exposed to the Sun for a longer period in each year than the Southern Hemisphere, must no doubt have a marked effect on the annual march of temperature. Even if the increased exposure is accepted as compensating for the weaker influence, it is evident that variations must result in the effects produced, from the variations in the cause of their production.

Although this is the case, the variations between the two hemispheres in regard to the annual march of temperature do not specially affect the subject of our present enquiry. It is clear that in both hemispheres, notwithstanding any points of difference, the temperature must be marked by an annual oscillation. The range from the minima to the maxima may differ, the duration of the upward or downward movement of the thermometer may be dissimilar, but, in each hemisphere, there will be a more or less regular upward movement during one period of the year, and a more or less regular downward movement at another period, these movements being the converse of each other in the two hemispheres.

Some authorities consider that, over the Northern Hemisphere as a whole, the temperature attains its maximum for the year in the month of August. The general consensus of meteorological opinion, however, is that the extreme months in the annual march of temperature are January and July. In January the Southern Hemisphere has its highest mean temperature, the Northern Hemisphere its lowest. In July the Northern Hemisphere has its highest mean temperature, the Southern Hemisphere its lowest. In many places in the Northern Hemisphere, no doubt, a higher temperature is attained in August than in July, and, possibly, in other places the maximum occurs in the end of June. On the average, however, it would seem to be the case that the yearly maximum in the temperature of the air occurs, in the Northern Hemisphere, in the latter part of July, and the yearly minimum in the latter part of January, the same periods coinciding, on the mean, with the yearly minimum and maximum respectively in the Southern Hemisphere. There is reason to suppose that in both hemispheres the temperature, generally speaking, reaches its maximum about a month after the occurrence of the summer solstice.

Thus the annual march of temperature, and the daily march of temperature, appear to have great similiarity in the relative periods of time which intervene between the Sun attaining its greatest power and the temperature reaching its highest level. In the daily oscillation of the temperature we may say that the interval is, on the mean, about two hours, which is the twelfth part of the solar day. In the yearly oscillation the interval would seem to be in general about one month, which is the twelfth part of the solar year.

The very evident similiarites which exist with reference to the occurrence of the daily maximum in the power of the Sun and the daily maximum of temperature on the one hand, and the yearly maximum in the power of the Sun and the yearly maximum of temperature on the other hand, suggest that the explanation of the occurrence of a time interval between these respective phenomena must be of the same character in each case. Let us now, therefore, endeavour to ascertain what the explanation really is.

The first explanation which might suggest itself is that the time interval which elapses between the Sun reaching the highest part of the heavens and the temperature attaining its maximum may be the time required for the heat-rays to pass from the Sun to the Earth. Such an explanation, however, is untenable. The speed of propagation of both light and heat is, as Professor Tait of Edinburgh has pointed out, "of the same order of magnitude for both phenomena, that is, 186,000 miles or so per second; for the Sun's heat ceases to be perceptible the moment an eclipse becomes total, and is perceived again the instant the edge of the Sun's disk is visible."

The transference of heat from one place or body to another occurs in three different ways, which are distinguished as radiation, convection, and conduction. Professor Tait illustrates these different modes of transference by what can be perceived to occur if a cannon-ball, which has been made red hot in a furnace, is experimented with. If the cannon-ball is suddenly removed from the furnace and suspended in the air, it at once begins to cool, or, in other words, to part with its heat, and it continues to do so at a gradually diminishing rate, until it finally reaches the surrounding temperature. In this cooling operation, two processes are simultaneously at work. One of these leads to the loss of heat in all directions indifferently; the other leads to a special loss in a vertical direction upwards. A column of air, very irregularly heated by contact with the ball, rises in the colder and denser air around it.

The process by which heat is lost indifferently in all directions is called radiation. It is, in fact, the process by which the Sun transmits heat to the Earth, and by which, on a comparatively insignificant scale, our own atmosphere loses heat outwards into space. The process which results in the irregular heating of the column of air by contact with the cannon-ball is called convection. "These two processes are," as Professor Tait remarks, "entirely different in their nature, laws, and mechanism."

The third method by which a transference of heat takes place—conduction—is illustrated by plunging the red-hot cannon-ball into water, and removing it again just as it ceases to be luminous. It will, in a short period, become once again red-hot, and will then gradually cool by the processes above referred to. In this case the heat passes from the interior of the ball to the surface.

In conduction, heat is transferred from the warmer to the colder parts of two contiguous portions of matter; it may be (as in Professor Tait's illustration), from hotter to colder parts of the same body. The heat, in fact, is transmitted through, or by means of, *a conductor*. In convection, an irregularly heated fluid becomes hydro-statically unstable—or, as we may express it, loses its equilibrium—and each part carries its heat with it to its new position, the heat being primarily transferred by contact. In radiation, heat is lost equally from all parts of the free surface, being radiated into the surrounding air, or into space, and not immediately taken up by any other body. In fact radiant heat and light are identical.

Just as certain substances allow light to pass through them freely, so certain substances allow heat to pass through them and are unaffected, or but little affected, by its passage. Indeed, very frequently, the same substances are transparent to both light and heat. We may quote Professor Tait:—

"That a body cools in consequence of radiation is certain

that other bodies which absorb the radiation are thereby heated is also certain ; but it does not at all follow that what passes in the radiant form is heat. To return for a moment to the red-hot cannon ball. If, while the hand is held below it, a thick but dry plate of rock-salt is interposed between the ball and the hand, there is no perceptible diminution of warmth, and the temperature of the salt is not perceptibly raised by the radiation which passes through it. When a piece of clear ice is cut into the form of a large burning-glass it can be employed to inflame timber, by concentrating the Sun's rays, and the lens does the work nearly as rapidly as if it had been made of glass. It is certainly not what we ordinarily call 'heat' which can be transmitted under conditions like these. Radiation is undoubtedly a transference of energy, which was in the form commonly called heat in the radiating body, and becomes heat in a body which absorbs it ; but it is transformed as it leaves the first body, and retransformed when it is absorbed by the second."

Lord Kelvin refers to the same phenomena in the following passage :—

" When two bodies at different temperatures are separated by a transparent medium, such as air, or water, or glass, or ice, heat passes from the warmer to the colder irrespectively of the temperature of the intervening medium, except in so far as its transparency may in some slight degree be affected by the temperature. Thus the colder of the two bodies becomes actually heated above the temperature of the intervening medium if the warmer be kept above this temperature, and if heat is not otherwise drawn off from the colder body in greater quantity than the heat entering it from the warmer. This process of transference from one body to another body at a distance through an intervening medium is called radiation of heat."

As the word "transparent" is specially applied as a descriptive term to bodies which permit the free passage of light, the word "diathermanous" has been adopted as descriptive of the quality of bodies which are freely permeable to heat, although, indeed, as we have noticed, light and radiant heat are believed to be identical. Thus Professor Cleveland Abbe, the meteorologist of the U.S. Weather Bureau, states that "the air is extremely diathermanous, or transparent to the general transmission of radient heat." In a still greater degree the ether of space is diathermanous. The rays of heat appear to pass from the Sun to our planet without having any effect on the intervening space, at least in so far as increase of temperature is concerned.

It would thus appear that, as regards the heat directly radiated from the Sun, the temperature of our atmosphere is affected only to a limited extent. Professor Abbe states that certain wave-lengths, or kinds of heat, are entirely cut off by aqueous vapour, and others by carbonic acid gas, and that, in general, the atmosphere appears to be "more transparent to the short wave-lengths." In fact most diathermanous bodies are opaque, or *athermanous*, to certain kinds of heat or wave-lengths. Rock-salt transmits nearly all the heat which falls on it, and the heat is transmitted whether of great or small wave-length. Other bodies act towards heat as coloured glass does towards light. They have a certain power of selection. White or clear glass transmits light unaltered, but, as regards heat, it stops rays of low refrangibility, or the longer wave-lengths, and can consequently be used as a fire-screen; while it transmits rays of high refrangibility, or the shorter wave-lengths, and consequently screens but little, if at all, from the heat of the Sun.

In the ordinary state of the atmosphere—notwithstanding the fact that "the air is extremely diathermanous"—a considerable amount of the heat which is received by our planet from the Sun is absorbed by the atmosphere. Professor Abbe considers that, with the barometer at about 29·9 in., and the Sun shining in the zenith, the proportion of the total heat radiated from the Sun towards any terrestrial position, and which actually reaches the surface of the Earth, is from 50 to 80 per cent. of that received at the outer bounds of the atmosphere. When the rays are more oblique, or when haze, dust, or cloud interfere, the transmission is further diminished. In general, one-half of the heat received by the illuminated atmosphere is absorbed by it, leaving the other half to reach the surface, provided there be no intercepting clouds.

The height of the terrestrial atmosphere has been estimated by Dr. Buchan at between 120 miles and 200 miles. The late Professor E. Loomis, of Yale College, U. S. A., estimated the height at about 500 miles, while other authorities reckon that it is considerably less than the extent estimated by Dr. Buchan. In view of the difference of opinion, the mean of the two extremes specified by Dr. Buchan may be taken as fairly representing the general scientific estimate. This would make the height of the atmosphere about 160 miles.

Now, if we accept the opinion of Professor Abbe, that in general only about one half of the heat from the Sun, which reaches the outer bounds of the atmosphere, actually passes through the atmosphere to the surface of the Earth, we may take it that the remaining half of the heat received by our planet from the Sun is distributed through a height of about 160 miles of atmosphere. The distribution will, of course, not be regular throughout the different strata of the atmosphere, as the absorption of the heat will depend upon the physical characteristics of the different layers. Still, the whole heat absorbed in the downward course to the Earth's surface will be, we may suppose, 50 per cent of the amount received; and this will be distributed through about 160 miles' thickness of atmosphere.

The remaining 50 per cent of the heat which reaches our planet from the Sun makes its way to the surface of the Earth, some of it reaching the solid surface, some of it the water surface, the proportion which reaches each description of surface being according to the superficial area and the exposure to the Sun. As both the land surface and the water surface differ materially from the atmosphere, the reception of the heat which reaches these surfaces is also materially different. Dr. Buchan makes reference to the absorption of heat by these surfaces :—

"Of the Sun's rays, which arrive at the Earth's surface, those which fall on the land and solid bodies generally are wholly absorbed by the thin surface layer exposed to the heating rays, the temperature of which consequently rises. Whilst the temperature of the surface increases, a wave of heat is propagated downwards through the soil. The intensity of the daily wave of temperature rapidly lessens with the depth at a rate depending on the conductivity of the soil, until, at about four feet below the surface, it ceases to be measurable. . . . Altogether different is the influence of the

Sun's rays on water. In this case the Sun's heat is not all, indeed very far from all, arrested at the surface, but penetrates to a considerable depth. The depth to which the influence of the Sun is felt has been shown by the observations made during the cruise of the *Challenger* to be, roughly speaking, about 500 feet below the surface of the sea."

Professor H. L. Callendar, now of London, and formerly of Montreal, has investigated the subject of the conduction of heat downwards through the soil. He was able, in some localities, by means of a platinum thermometer, to measure the daily temperature wave at a greater depth than four feet. At Montreal, indeed, he obtained a measureable variation at a depth somewhat greater than eight feet. He states, however, that, with a daily range of temperature at the surface of about 20 degrees Fah., the range at a depth of only four inches was reduced to about six degrees Fah., and, at a depth of about ten inches, to less than two degrees Fah., the epochs, or time of occurrence of maximum and minimum temperature being also greatly changed. In his opinion, the percolation of water has much to do with the diffusion of heat to the greater depths. It may, therefore, be taken as approximately correct that at a depth of about four feet below the surface, the daily wave of temperature, generally speaking, ceases to be measurable.

If, now, we suppose a certain amount of heat to reach our planet from the Sun immediately above a certain area of water, we may take it, broadly, that one-half of the heat is absorbed by 160 miles depth of atmosphere, and the other half by 500 feet depth of water. If, on the other hand, the heat is received above a land surface, we have, as in the former case, one-half of it absorbed by the 160 miles of atmosphere, whilst the other half is absorbed by about four feet thickness of land surface. In every case the absorption will be irregular, and, in the case of the land and water surfaces, will depend materially on the distance of the respective strata from the surface level.

If, then, the same amount of heat is absorbed by 500 feet of water as is absorbed by 160 miles of atmosphere—the superficial area at the surface level being in both cases the same—the water must, bulk for bulk, receive much more heat than the air. The same comparison may be made between the land surface and the water surface. We have four feet depth of land absorbing as much heat as 500 feet depth of water. Thus, in the case of the land, the heat is concentrated in a very small bulk; in the case of the water it is more widely distributed; while, in the case of the atmosphere, it is excessively wide-spread. In fact, as regards the atmosphere, the distribution is even greater than we have indicated, as the superficial area of the different strata will, according to the distance from the surface level, be greater than the superficial area of the land and water surfaces at the surface level.

From this it follows that by the influence of solar radiation the land surface is *most heated*; the atmosphere *very slightly heated;* and that the oceans occupy an intermediate position, being warmed up very much more than the atmosphere, but distinctly less than the land.

Now, as Lord Kelvin points out, "when two contiguous portions of matter are at different temperatures, heat is transferred from the warmer to the colder." Thus, as the surface of the Earth, whether land or water, is more heated by the influence of solar radiation than is the atmosphere which rests upon it, heat must necessarily pass from the surface to the atmosphere; and, as the land is more heated than the water, so the air in contact with the land will receive more heat than the air in contact with the water. It would, therefore, appear that the main cause of the warming of the atmosphere is not solar radiation, but terrestrial convection.

As Dr. Buchan states:—

"Part of the heat of the surface layer is conveyed upwards through the air by the convection currents which have their origin in the heating of the lowermost stratum of air in direct contact with the heated surface of the land."

The same matter is succinctly referred to by the author of the article on "Heat" in Tomlinson's Cyclopædia:—

"The atmosphere receives scarcely any of its warmth directly from the sun's rays, but is heated almost entirely by the ground on which it rests, and is, therefore, in the condition of the water in a boiler, where the heat is applied from below.'

This, then, seems to be the explanation of how the maximum temperature, whether for the day or year, occurs only a considerable time after the Sun has passed its highest point in the heavens, notwithstanding the fact that heat is propagated with the velocity of light. The terrestrial surface, no doubt, receives the heat waves from the Sun as quickly as it receives the waves of light, but the atmosphere only slowly takes up the heat from the surface, and time is required for its diffusion.

In the relation between the *daily* progress of the Sun

and the daily march of temperature, the applicability of such an explanation is readily perceptible, but it is a little more difficult to grasp its applicability to the *annual* phenomena. The difficulty, however, disappears if the close analogy between the year and the day is kept fully in mind. The daily temperature oscillation is indeed an insignificant movement in comparison with the annual temperature oscillation. The former may be compared to the waves or undulations of the sea in their passage shorewards, the latter to the tidal movement in its periodic rise and fall.

Day after day, during one period of the year, we have a gradual rise of temperature, masked though it may be to a trifling extent by the daily oscillation. Day after day, during another period of the year, we have a gradual fall of temperature, although it also is characterized by the daily rise and fall. During the former period, the hemisphere concerned receives on the average more heat during each diurnal rotation than it loses, and the movement of the thermometer, although subject to a diurnal oscillation, is more or less steadily upwards. During the latter period, the converse is the case. Thus we might describe the *daily* rise and fall of temperature to be on the whole of a fairly regular character, and the *annual* rise and fall to be of an oscillatory or wavy character.

It is evident that, in our individual experience, a variation of temperature, as influenced by the Sun, arises in two entirely different ways. We may be warmed by the direct influence of solar radiation, as occurs when we expose ourselves to the sunshine; or we may be warmed indirectly by the Sun, through the increasing temperature of the air as it becomes gradually heated. The first of these methods is that referred to in Professor Tait's statement that "the sun's heat ceases to be perceptible the moment an eclipse becomes total, and is perceived again the instant the edge of the sun's disk is visible." The second has reference to the temperature of the atmosphere in the shade. These different ways in which we experience the warming influence of the Sun, are illustrated by the noticeable difference of temperature between sunshine and shade, which is of frequent occurrence in the spring and early summer in the high temperate latitudes. It is, of course, evident that in the observation of the daily and annual march of temperature, and the fixing of the times of the respective maxima, it is the temperature of the atmosphere in a position where the thermometer is sheltered from the effect of radiation which is recorded.

From all this, it follows that the temperature of the air usually lessens as the distance from the surface is increased. During the years from 1862 to 1866, a committee of the British Association investigated, amongst other atmospheric phenomena, the subject of change of temperature in its relation to distance from the Earth's surface. Mr. James Glaisher, F.R.S., a member of the committee, made a number of balloon ascents in connection with the enquiry. On 18th August, 1862, an ascent was made from Wolverhampton, when the temperature near the surface was 67·8° F. As the balloon rose, it was found that the temperature of the air through which it was passing, was steadily lessening. At the height of 11,500 feet the temperature was 39·5° F., and the balloon then descended to about 3000 feet above the surface, and the temperature gradually increased to 56° F.

and the daily march e temperature, the applic
such as explanation s readily perceptible, b
little more difficult t grasp its applicabilit
annual phenomena. he difficulty, however, d
if the close analogy tween the year and th
kept fully in mind. he daily temperature c
is indeed an insignificant movement in compar
the annual temperature oscillation. The form
compared to the waveor undulations of the se
passage shorewards, th latter to the tidal mo
its periodic rise and fa

Day after day, durin one period of the ye
a gradual rise of temperature, masked though
to a trifling extent by he daily oscillation.
day, during another period of the year, we ha
fall of temperature, although it also is char
the daily rise and fall. During the former
hemisphere concerned ceives on the avera
during each diurnal rotation than it loses, a
ment of the thermometr, although subject
oscillation, is more or less steadily upward
latter period, the converse is the case.
describe the *daily* rise nd fall of tempe
the whole of a fairly regular character
rise and fall to be of an

It is evident that
variation of tem
in two enti
the direct
expose
indirect
of the

these methods is that referred ... Professor ...
statement that " the sun's heat ... is ...
the moment an eclipse becomes ... and ...
again the instant the edge of the sun's disc is visible."
The second has reference to the temperature of the
atmosphere in the shade. That ... says a text ...
we experience the warming influence of the sun, as
illustrated by the noticeable difference of temperature
between sunshine and shade, which is of frequent occur-
rence in the spring and early summer in the ...
temperate latitudes. It is, of course, evident that in the
observation of the daily and annual march of temperature
and the fixing of the times of the ... maxima, the
the temperature of the atmosphere in a position ...
the thermometer is sheltered from the ... of ...
which is recorded.

From all this, it follows that ... temperature of the ... air
air usually lessens as the distance from its surface is ...
increased. During the years ... 1862 to 1869, a ...
committee of the British Association ...
amongst other atmospheric phenomena the ... and the
change of temperature in its relation to distance from
the Earth's surface. Mr. Jas. ... the ...
member of the committee, had ... investigation, came
... in ... tion with the ... ... ombay, the
... de from ... Madras, from
... nd will come
region of equal
... mmer, the daily
... of the wind, be
... while, in Madras,
... effect.

... rface, and

Ballast was thrown out, and the balloon rose to 23,000 feet, the temperature falling, pretty steadily, to 24° F.

On 5th September, 1862, a remarkable ascent was made from the same city, the height attained being, it was supposed, over six miles. The temperature near the surface was 59° F., and, with each mile of ascent, it was found to be as follows: at one mile 41° F., at two miles 32° F., at three miles 18° F., at four miles 8° F., at five miles 2° below zero. Shortly after this elevation was attained, Mr. Glaisher temporarily lost consciousness, evidently through the extreme rarity of the air, and his companion also was partially incapacitated. The latter, however, was able, with much difficulty, to open the valve with his teeth, and thereby caused the balloon to descend. The height reached on this occasion appears to have been the greatest attained by any aeronaut, at any rate up to 1901. On 31st July, 1901, Dr. Berson and Dr. Süring were able, by imbibing oxygen, to ascend from Berlin to a height of about 34,000 feet—which is nearly 6½ miles—being about the same height as is supposed to have been attained by Mr. Glaisher during his period of unconsciousness.

On 17th July, 1862, Mr. Glaisher had experience of a warm current of air at a great elevation. He ascended from Wolverhampton at 9.43 a.m., the temperature being 59° F. The temperature gradually fell, until the balloon was at a height of 10,000 feet, the temperatnre at that height being 26° F. From 10,000 feet to 13,000 feet, the temperature was stationary, and it then began to increase, until, at 19,500 feet, it was found to be 42° F. It then rapidly fell and at 26,000 feet it was 16° F. On descending, however, a regular increase of temperature was experienced. Mr. Glaisher expresses the opinion that "the rise of temperature met with in this ascent was most remarkable."

There would seem thus to be no doubt that the main factor in the heating of the atmosphere is terrestrial convection, and that the temperature of the air, as a general rule, becomes less as the separation from the surface is increased. Consequently, the occurrence of a time interval, between the maximum power of the Sun and the maximum height of the thermometer, arises from the fact that the heating of the air through contact with the surface, and the gradual diffusion of the heat, is necessarily a comparatively slow process.

It will be noticed from the tables we have given, that the time when the atmosphere attains its maximum temperature varies in different localities as regards both the daily and annual extremes, and that it also varies in the same locality in different seasons. Let us shortly consider these peculiarities, dealing first with the variation of time in different places.

It is clear that the effect on the surface of solar radiation will vary according to the nature of the surface at which the heat is received. We have already noticed that the effect is very different according as the surface is land or water. The effect will also be different according to the character of the land or water. A land surface may be rocky, sandy, clayey, or loamy, or it may have no pronounced character. Its state as to moisture may also vary. The absorption of heat and its communication to the air will differ accordingly. As Dr. Buchan puts it:—"The intensity of the daily wave of temperature rapidly lessens with the depth at a rate

depending on the conductivity of the soil." In some
cases the heat will be quickly communicated to the
atmosphere and quickly diffused. In others, it will be
slowly and gradually passed on to the atmosphere. In
a less degree, a corresponding difference will be occasioned
by the character and varying density of water surfaces.
Other local conditions, also, will have an effect in the
production of the variations in different localities. We
have noticed that Dr. Buchan explains the occurrence of
the daily maximum temperature "so early as noon," at
certain places visited by the *Challenger* near land, as
being "doubtless occasioned by the greater strength of
the sea breeze after this hour." Thus local winds have
probably considerable effect in producing local variation
in the hour at which the temperature for the day reaches
its highest point.

The seasonal variations in the time of maximum atmospheric temperature in the same locality are of much
interest. The table on page 29 shows the magnitude of
these time variations in so far as they occur at the fourteen places mentioned in the preceding table.

It appears that, out of these fourteen stations, which
may be accepted as fairly representative, the maximum
temperature occurs later in summer than in winter at all
but three. What speciality distinguishes these three
from the other stations? Let us take a glance at Bombay
and Madras, which are both tropical stations, and in
which the variations are almost diametrically opposed.

In Bombay, the daily maximum occurs 40 minutes
earlier in summer than in winter, while, in Madras, it
occurs 45 minutes earlier in winter than in summer.
Both cities are on the coast, so that mere proximity to

the sea can have no differentiating effect between them. It seems not improbable that the difference is really occasioned by the wind. Both Madras and Bombay are exposed to the strong south-west monsoons in the summer months. As Bombay is situated on the west coast, and Madras on the east, these winds will, although

TABLE SHOWING THE MAGNITUDE OF THE TIME—DIFFERENCE IN THE OCCURRENCE OF THE DAILY MAXIMUM TEMPERATURE IN SUMMER AND WINTER, AT CERTAIN STATIONS IN THE NORTHERN HEMISPHERE.

Town.	Seasonal Difference in time.		Town.	Seasonal Difference in time.	
	Earlier in Summer.	Earlier in Winter.		Earlier in Summer.	Earlier in Winter.
	Minutes.	Minutes.		Minutes.	Minutes.
Archangel		80	Toronto		115
Bogoslovsk		35	Tiflis		45
Sitka	40		Madrid		10
Rothesay		45	Philadelphia		30
Oxford		70	Calcutta	110	
Geneva		50	Bombay	40	
St. Bernard		25	Madras		45

in both cases they blow from the same direction, come from a different character of surface. In Bombay, the wind will blow from the sea landwards; in Madras, from the land seawards. In the former, the wind will come from a cooler region; in the latter from a region of equal or greater heat. Thus, in Bombay, in summer, the daily maximum may, by the cooling influence of the wind, be reached more quickly than in winter; while, in Madras, the same wind may have an opposite effect.

It appears that in Calcutta, as in Bombay, the maximum temperature for the day occurs earlier in summer than in winter. In fact, in Calcutta, the time of occurrence is one hour fifty minutes earlier in summer, while in Bombay it is only forty minutes. It would seem that, as regards the character of the neighbouring surface from which the south-west monsoon of the summer months will arrive, Calcutta is situated similarly to Bombay. At both places the monsoon will be a sea-wind. In Bombay the monsoon will blow from the Arabian Sea, while in Calcutta it will blow from the Bay of Bengal. In each case, therefore, the same cause probably contributes to the occurrence of the peculiarity which distinguishes these two cities from the other tropical stations mentioned in the table, as regards the seasonal variation in the time of occurrence of the daily maximum temperature.

Although Sitka, in Alaska, is the only other station mentioned in our Table, in which, as in Calcutta and Bombay, the maximum temperature occurs earlier in the day in summer than in winter, this, no doubt, arises merely from the limited number of stations with which we are dealing. These three stations must be accepted as simply representative. In like manner the explanation which we have suggested as applicable to Calcutta and Bombay is representative of the explanations of the differences in other stations. Local geographical and climatic conditions, in their bearing on the changing seasons, would seem to be the deciding influences, as regards both the extent, and the direction, of the variation in time of the daily maximum throughout the year in the same station.

The more general prevalence of a later occurrence of the maximum in summer than in winter, brought out in the table, is most likely correct, notwithstanding the limited number of places with which we have dealt. Indeed, it would seem reasonable to expect that, unless other causes counteract such a result, the maximum temperature should occur later in the days of summer than in the days of winter. In summer the surface is heated to a much greater extent than it is in winter, and thus a greater quantity of heat has daily to be transferred from the surface to the atmosphere, and diffused throughout the atmosphere. Other conditions being similar, the transference and diffusion of a greater amount of heat must occupy a longer time than the transference and diffusion of a smaller amount. Out of the fourteen stations mentioned in the table, the maximum temperature is attained later in summer at no less than eleven, the time by which it is later varying from ten minutes in Madrid to nearly two hours in Toronto. No doubt the extent of the variation is decided by geographical and meteorological conditions of a purely local character.

We might, under our subject "The Turn of the Day," deal also with the converse "turning" movements which characterize the solar day and the solar year. These occur when the Sun at midnight is farthest below our horizon in the diurnal rotation; and when, at midnight, at the time of the winter solstice, the Sun is most completely absent from us. We have, however, in both the diurnal and the annual aspects of our subject, restricted ourselves to dealing with the movements in their relation to the "day" as distinguished from the night, the period of light as distinguished from the period of darkness.

It is noticeable that, in contrasting the periods of the maximum and minimum power of the Sun with the times of occurrence of the maximum and minimum daily temperature, two distinct meanings of the terms "summer" and "winter," respectively, are apparent. Neither of these meanings can be considered as inappropriate, provided the sense in which either is used is made clear.

If these seasons are considered in relation to the power of the Sun, then, in these latitudes, we should evidently have midsummer when the Sun reaches its most northern declination, and midwinter when it is farthest south. This gives us midsummer at about the 21st of June, and midwinter at about the 22nd of December. Indeed, if we speak of midsummer or midwinter, in the Northern Hemisphere, these dates, or neighbouring dates, are, of course, understood.

If, on the other hand, we consider the seasons, summer and winter, in relation to temperature, summer being the warmest season of the year and winter the coldest, we certainly cannot accept the dates mentioned as constituting, with any degree of accuracy, in these northern latitudes, the middle of summer or the middle of winter. Both would have to be fixed about a month later.

There is yet a third meaning attached to the terms "summer" and "winter," which has not been made evident in our present subject, a meaning which also has strict reference to the Sun. Astronomically considered, summer commences in the Northern Hemisphere when the Sun enters Cancer at the June solstice; and winter commences when the Sun enters Capricornus at the December solstice. This meaning in itself is indefinite, as it has now no direct relation to the constellations of these names, and merely bears reference to the distance of the Sun from its place at the vernal equinox, "the first point of Aries," a position which is slightly different in its situation every succeeding year.

Thus the word "summer" is, in these northern latitudes, used indifferently to signify (1) a season of, say three months' duration, of which the middle date is about the 21st of June; (2) a season of the same duration, of which the middle date is about a month later than the 21st of June; and (3) a similar period, *commencing* on or about the 21st of June. The same statement would apply to the word "winter," if we substitute the 22nd of December for the 21st of June. All three meanings are undoubtedly thoroughly defensible, and yet the fact that these different meanings have each scientific acceptance is calculated to occasion ambiguity.

It has been said of the late Sir Richard Owen, the celebrated biologist, that he could, if supplied with a single bone of an extinct animal, reconstruct the whole skeleton. Such a statement may be an exaggeration, but yet it illustrates the value in scientific investigation of the apparently unimportant details. We have endeavoured to deal with what may appear to be a comparatively small matter in relation to the science of meteorology, and we have found that it leads imperceptibly to other matters of importance. It would seem as if the complete investigation of the relations between the place of the Sun and the annual and daily march of temperature would result in a knowledge not only of the local variations in the state of the atmosphere, but of the local variations in the nature of the soil. It is thus difficult to set precise limits to such a subject, or to judge as to the importance of its bearing on the science of meteorology as a whole. There can, however, be no doubt that, even in connection with such a trivial occurrence as the daily and yearly "Turn of the Day," there are many details worthy of careful investigation, the full understanding of which would materially advance our knowledge of the laws applicable to the seasons in their varied geographical relations.

THE NORTH POLE AND ITS PECULIARITIES.

D

SYNOPSIS.

Early polar exploration—Reward offered by Parliament—Derivation of term " Pole "—Nearest approach to North Pole—Scientific and geographical results achieved—Distinguishing features at North Pole—Stellar aspect—Aspect of Planets and Sun and Moon—Aurora—Periods of light and darkness—Sun's apparent daily motion—Eclipses—Precession of the Equinoxes—Nutation—Temperature—Ocean currents—Winds—Force of gravitation—Time reckoning—Mean length of solar day shorter than elsewhere—Sidereal clock—Length of year—Atmospheric pressure—Polar ocean—Polar inhabitants and animal and plant life—Oscillation of Earth's axis—Past and future exploration.

THE NORTH POLE AND ITS PECULIARITIES.

For at least a century and a half the discovery of the North Pole has been the ambition of explorers, and a subject of perennial interest to scientists.

In February 1773, the Royal Society of London submitted a proposal to King George the Third for sending an expedition to try how far navigation was possible towards the Pole. Two ships, the *Racehorse* and the *Carcass*, were fitted out for the purpose, and they sailed on the 2nd of June 1773. This appears to have been the first regularly organized British expedition in the interest of polar exploration. It succeeded in reaching latitude 80° 48′ N. The Russians had, however, been previously engaged in the same work, and they, during the 18th century, were able to delineate the whole of the northern coast of Siberia. One Russian explorer—Count Vassili Tchitschakoff—penetrated, about 1766, as far north as latitude 80° 30′, being rather more than 20 miles short of the latitude attained by the expedition organized by the Royal Society.

In 1776 the British Parliament passed an Act offering a reward of £5000 to anyone who should approach by sea within one degree of the North Pole. This Act continued in force till 1828, when it was repealed. Although more than three quarters of a century have now passed since the repeal of the Act—a period marked by most strenuous exertion on the part of polar explorers—the conditions attached to the proposed reward have never yet been fulfilled.

The name "Pole," as applied to the ends of the Earth's axis in the terms "North Pole" and "South Pole," has, of course, no relation to the ordinary use of the word. It is derived from the Greek *polos*, which means a *pivot*, or *axis* on which something turns, particularly the axis of the globe. In our use of the word as a geographical term, we restrict its application specially to the *extremities* of the axis. The word "pole," in the sense of a long piece of timber, has a totally different derivation.

The nearest approach which has yet been made to the North Pole was made by a sledge party from an expedition carried out by Prince Luigi, Duke of the Abruzzi, a cousin of the King of Italy. His ship, the *Stella Polare*, put into Teplitz Bay, in Rudolph Land, to pass the winter, 1899-1900. Three sledge parties were sent north in the spring of 1900, and one of these, under Captain Cagni, reached latitude 86° 33' N., in longitude about 56° E. The minimum distance from the Pole was therefore 3 degrees 27 minutes, which, if we take the degree of latitude to be equal to 69 miles, would represent about 238 miles. This is about the distance which separates London from the Lake District of the North of England, or about 34 miles less than the greatest length of Scotland from north to south. The farthest north attained by Nansen, in 1895, was 86° 13' 6", which was 19' 54", or about 23 miles, short of Cagni's record.

SKETCH MAP OF THE ARCTIC REGIONS.

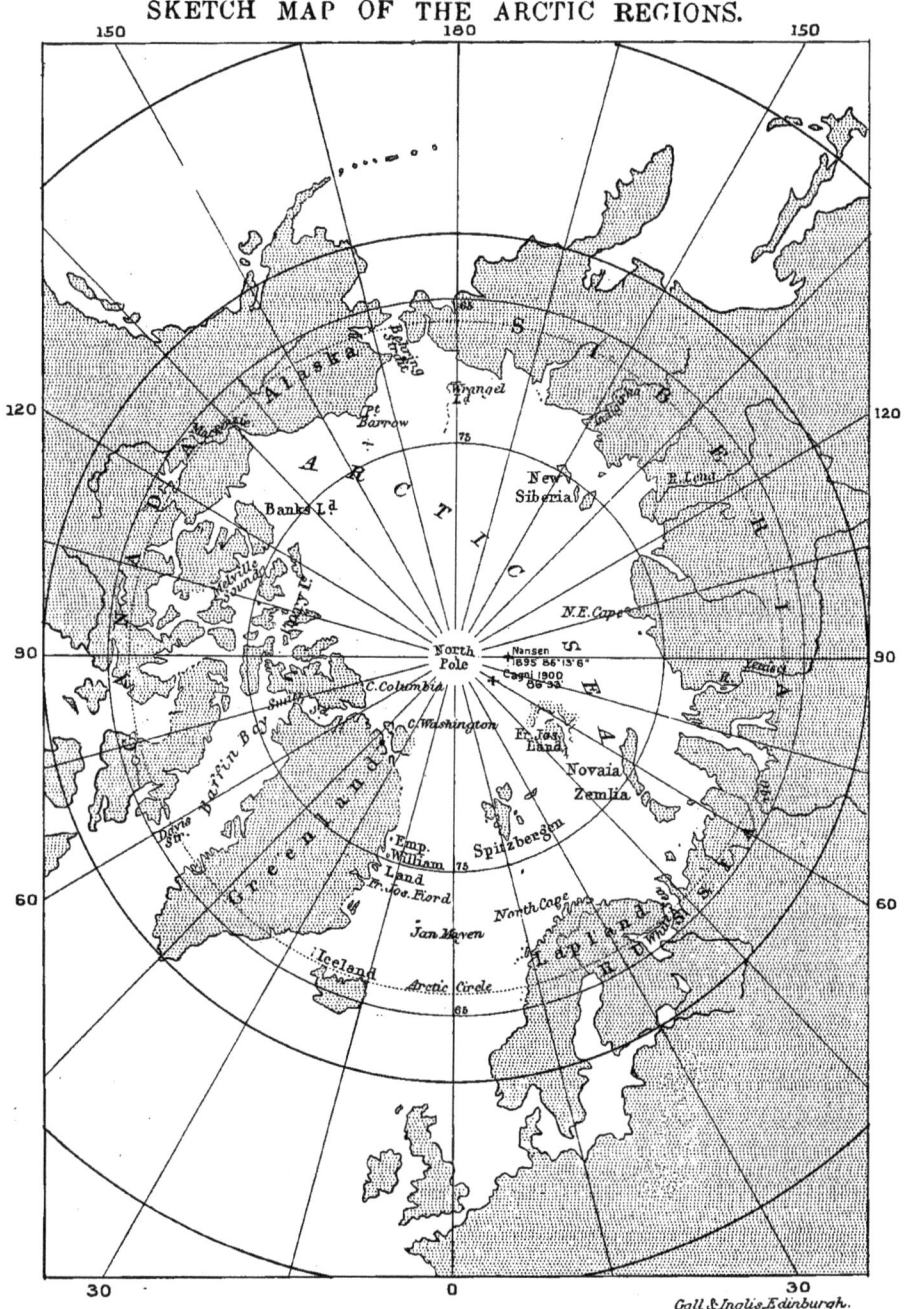

Gall & Inglis, Edinburgh.

Although the attempts to reach the North Pole, laborious and carefully planned as they have been, have hitherto failed to achieve their primary purpose, they have by no means been devoid of success. In fact, the advantages which science and geographical knowledge might be expected to derive from the discovery of the Pole, have been, in many respects, secured through the polar exploration which has been carried out in the effort to reach it. Like the farmer who rose early to get a sight of the white sparrow, and who thereby, although he never saw the white sparrow, gained knowledge for the guidance of his own affairs, every North Pole expedition, although unsuccessful in reaching its destination, has added to our knowledge of the conditions prevailing at the Pole. In fact it has now come to be recognised that the North Pole has, for science, no greater significance than any other point in the higher latitudes. We have, through scientific investigation and polar exploration, acquired so much information regarding the conditions applicable to the North Pole, as to have, to a great extent, anticipated detailed inspection, and this although no civilized man has ever yet reached that position, or, at any rate, ever returned to report his having done so.

Let us now consider what peculiarities, if any, distinguish the North Pole from other positions on the surface of the Earth.

One of the chief distinguishing features of the Pole is in the aspect of the heavens. Near the zenith is *Polaris*, the Pole-star, its distance from the zenith being less than a degree and a quarter, or say about twice the width of the Moon's disk. Of the visible or bright stars, it is the nearest to the zenith, and is conspicuous as it circles

steadily around the highest point in the heavens. All
the other stars in the celestial vault describe similar
circles, every circle having the zenith as its centre. Thus
the Pole-star, being the nearest to the zenith, describes
the smallest circle, and the remaining stars, according to
their distance from the zenith, describe larger circles. In
the lower latitudes, we are accustomed to see stars rising
in the east, passing westwards, and sinking below the
western horizon. This does not occur at the North Pole.
The stars which are now in view remain in view. They
do not rise, and they do not set. They simply and constantly pass round and round in unending circles.

Although the Pole-star is near the zenith, and the circle
which it describes is therefore a small one, it takes exactly the same time to complete its circle as the stars which
are lower in the sky. The stars which are just above
the horizon, and which have, consequently, to describe an
immense circle, revolve in exactly the same time as the
Pole-star itself. The apparent motion of the stars will
thus appear to vary greatly. The Pole-star will seem to
move very slowly, while the stars on the horizon will
seem to move very quickly. The movement is similar to
the revolving of a wheel, the zenith representing the
axle. Near the axle is *Polaris*, while near the rim of
the wheel are the bright stars *Betelgeux* in Orion, *Procyon*
in Canis Minor, *Regulus* in Leo, and *Altair* in Aquila.
The direction of the circulating movement, as viewed by
an observer at the Pole, is from left to right, and thus
is similar to the movement of the hands of a watch or
clock, as seen from the centre of the dial around which
they revolve.

The time taken by every star to complete its circle

Chart of the principal stars visible at the North Pole.—p. 40.

around the centre of the heavens is 23 hours, 56 minutes, 4·1 seconds, being the time known as the sidereal day. As the movement of the stars in their circle around the zenith is of course only apparent, and is occasioned by the rotation of the Earth, the time mentioned is the actual time of the Earth's diurnal rotation.

The stars which, at the North Pole, would be seen on and near the horizon, are the stars which, at the equator, appear to pass from east to west directly through the zenith. As the most northerly star in the belt of Orion is virtually on the celestial equator, it would, at the North Pole, be just visible on the horizon. Probably, through refraction, stars slightly south of the equator may be observable, and thus the whole belt of Orion may perhaps be perceptible. A resident at the North Pole would, however, have no conception of the general outline of that magnificent constellation. Of the twelve constellations in the Zodiac, at least five are unknown at the North Pole—Libra, Scorpio, Saggitarius, Capricornus, and Aquarius. *Sirius*, the brightest star in the firmament, has no existence, so far as any observer at the North Pole is concerned. Of all the stars north of the celestial equator, however, such an observer would have grand opportunities for observation, and these would include at least the greater part of each of the remaining constellations of the Zodiac. In the long night of the North Pole, all these stars would be constantly in view. In the clear atmosphere of the cold regions, their sparkling brilliance would necessarily attract observation. Were it possible for human beings to live at the North Pole, we might conclude that, unless they hibernated for six months of the year, one and all would have a fair knowledge of astronomy, in so far as the Northern Celestial Hemisphere is concerned.

Ptolemy, in the second century of our era, taught that the Earth was the centre of the universe, around which the stars revolved. This idea might most naturally be accepted at the North Pole, were the evidence of the senses believed. Every star in the heavens will appear to circle daily around the habitation of any residents in that part of our globe, the North Pole being actually the terrestrial centre of the circle described by every star.

While the stars thus, to an observer at the North Pole, will appear to describe regular concentric circles, there are certain apparent stars which will have a different kind of motion. These are, of course, the planets. They will appear on the horizon at uncertain seasons, and will then circle spirally around the heavens in an upward direction, mounting for a time higher and higher, till they are between 20 and 30 degrees above the horizon, and then gradually, by spiral circles, descending again until they once more disappear below the horizon. This spiral movement will, at the North Pole, distinguish a planet from a star, and, with the stars continually in view for six months at a time, should render the recognition of a planet a matter of much less difficulty than it is in lower latitudes.

The movement of the Sun and of the Moon will be of the same character as the movement of the planets. They will appear on the horizon, and will circle around the heavens in a spiral direction, mounting slowly upwards till they attain their maximum elevation, and then as slowly descending, while they continue to circle in the same direction. Throughout the long winter, when the

Sun is hidden from view, the full Moon will every month reach a fair elevation above the horizon, the maximum height attained by the Moon being rather less than 20 degrees. The Moon will thus, for a week or more at a time, continuously shed its pale radiance over the gloom of the Arctic area.

The Moon, as it circles spirally around the heavens, will be seen to fall behind the stars, as if its rate of movement were slower. If it is in conjunction with a star, when it appears above the horizon, it will be found on completing its spiral circuit to be more than 13 degrees behind it, although, of course, at a higher level. This difference is occasioned by the real motion of the Moon, as distinguished from the apparent motion arising from the Earth's rotation. The real motion is therefore in the reverse direction from that in which the Moon appears through the Earth's rotation to be moving. This motion of the Moon relative to the stars will be very perceptible at the North Pole, and will appear to be owing to its movement being slower than that of the stars, as if its energy were absorbed by its transition upwards and downwards. For about a week after it rises above the horizon, the Moon will continue to mount upwards, and, for about a week afterwards, its movement will be downwards, till at length it is again lost at the horizon, from which it will once more arise about fourteen days later.

The aurora will be very noticeable in the heavens during the winter darkness at the North Pole, although there does not seem to be any reason to suppose that it will be more conspicuous there than in other parts of the polar regions. It is thought that the aurora is connected with terrestrial magnetism, and, consequently, that

auroral displays are more centralized in the neighbourhood of the magnetic poles than at the geographical
poles. As the northern centre of magnetic attraction is
in high latitudes, the aurora will, no doubt, at the North
Pole be exceedingly prominent. If the centre of its
attraction, however, is the magnetic pole, it may be conjectured that, at the North Pole, it will not conform to
the general stellar movement of describing a circle around
the polar zenith; still, it will certainly, at the Pole, not
infrequently mitigate the darkness of the long Arctic
night.

At the equator, the Sun is in the zenith at the vernal
and autumnal equinoxes, which occur on or about 21st
March and 23rd September annually. Therefore it might
naturally be supposed that, on these dates, the centre of
the Sun should be exactly on the horizon at the North
Pole, one half of the disk being above, and the other half
below the horizon. This would be the case but for two
circumstances, which each cause a slight variation.
Owing to the vast size of the Sun in comparison with the
Earth, slightly more than one half of the Earth is em-
braced at the same time in the direct flow of light from
the Sun. The excess over one half of the Earth so em-
braced, is calculated to amount to 16 minutes of the arc
of the circle. The second circumstance is that—owing
to the refraction of the Sun's rays in passing through the
Earth's atmosphere—the Sun, when near the horizon, is
apparently thrown upwards, so that, while still below
the horizon, it is visible as if above it. Through refraction, the light of the Sun is calculated to cover 35
minutes of the arc of the circle of the terrestrial hemisphere, in which, otherwise, the Sun would be below the

horizon. These two facts will result in the North Pole being exposed to the Sun about a couple of days before the vernal equinox, and about a couple of days after the autumnal equinox.

From about the 25th of September to about the 19th of March the Sun will be under the horizon at the North Pole. It must not be supposed, however, that the Pole will, so far as sunlight is concerned, be in darkness during the whole of that period. Although the Sun will disappear from view a day or two after the autumnal equinox, it will be circling around immediately below the horizon, and its further descent is very gradual. Consequently, twilight at the Pole will continue for weeks after the Sun has disappeared. When the Sun's declination is 9 degrees S., which does not occur until about the 16th or 17th of October, the North Pole is in the centre of the twilight belt, and, consequently, the last trace of twilight will not disappear until a further corresponding interval after the autumnal equinox. This would bring us to the second week in November. A similar period of gradually increasing twilight will occur before the vernal equinox. It is estimated that the actual period during which the North Pole is in total darkness, in so far as the light of the Sun is concerned, is about two and a half months.

When the Sun, at the vernal equinox, appears above the horizon at the North Pole, it will circulate spirally upwards at the rate of about a quarter of a degree in each revolution. This will continue until the summer solstice, which occurs about 21st June, and the Sun will then have an elevation of about $23\frac{1}{2}$ degrees above the horizon. It will thereafter slowly descend in a similar manner, until it sinks below the horizon shortly after the autumnal equinox. Thus, for rather more than six months annually, the North Pole is continuously exposed to sunshine, unless atmospheric causes intervene.

The Sun, like the Moon, will take a longer time to complete the circuit of the heavens than is taken by the stars, the excess being the slight amount by which the solar day is longer than the sidereal day, which, on the yearly mean, is about 3 minutes 56 seconds. This, of course, arises from the orbital motion of the Earth, which has the effect of displacing the Sun in the opposite direction from that in which it appears to move owing to the Earth's daily rotation. The Sun's apparent motion at the North Pole will be very little less than that of the stars. In each revolution around the heavens, the Sun will appear not only to have changed its height above the horizon by about a quarter of a degree, but to have fallen about one degree behind any star which commenced the revolution at the same point.

Eclipses of the Sun and the Moon occur at the North Pole as in other latitudes, but, of course, either the Sun or the Moon must be north of the celestial equator for either a solar or lunar eclipse respectively to be perceptible. The total eclipse of the Sun which occurred on 30th August, 1905, and which, at Greenwich, obscured ·786 of the Sun's disk, was visible as a partial eclipse of considerably less extent at the North Pole. As the Moon is actually deprived of its light during a lunar eclipse, every person on the face of the Earth who sees the Moon, sees the eclipse. The North Pole, during the Arctic winter, would evidently be a peculiarly well-situated position for the observation of lunar eclipses.

We have noticed that, at the North Pole, the stars will all appear to move around the heavens in concentric circles, the centre of which is situated in the zenith. This statement is, however, subject to a slight qualification. Owing to what is known as "the precession of the equinoxes" the part of the heavens occupying the zenith at the North Pole is not absolutely constant. The North Pole has a gyratory motion, whereby it describes a circle, in the course of ages, on the part of the heavens towards which it points—the diameter of the circle being about 47 degrees. This precessional movement of the Pole is, however, excessively slow. It amounts to about 50·1 seconds of arc in a year, and thus it will take about 25,868 years to complete the circle. This movement is at present in such a direction that the North Pole is yearly pointing nearer to the Pole-star by an almost imperceptible distance. In about 200 years, the part of the heavens occupying the zenith at the North Pole will be less than half a degree distant from *Polaris*. This gyratory movement of the North Pole will have the effect of bringing into view certain stars, on the side towards which the movement for the time may be directed, which were formerly below the horizon, and raising the horizon on the other side and obscuring certain stars which were previously visible. As the movement of the stars is only apparent, this displacement of the zenith will cause a corresponding displacement in each of the concentric circles described by the stars in their apparent revolution, the circles, of course, remaining concentric.

The movement of precession, whereby the North Pole describes a circle in the heavens by the constant shifting of the point towards which it is directed, is somewhat complicated by another movement of the Earth's axis, which is known as " Nutation." The nutational movement of the axis results in the circumference of the circle described in the precessional period being a series of minute waves, instead of a uniform line. Each wave occupies about 18⅔ years, so that the circumference of the circle traced in the precessional period of 25,868 years will consist of nearly 1,400 waves. Both these movements are believed to be due to the varying attractive influence of the Sun and Moon on the spheroidal figure of the Earth. It is evident that, as each results in a displacement of the point occupying the zenith at the North Pole, they will cause a corresponding displacement in the stars which move around the zenith-point as a centre. The displacement is, however, so exceedingly slow that it will only be perceptible by observation extending through a prolonged period. If we conceive of the North Pole as an inhabited part of the Earth, we may imagine the aged residents drawing attention to the fact that certain changes have occurred during their time in the appearance of the heavens, that stars which were visible at the horizon in their youth have disappeared and other stars have come into view, while the circle described by *Polaris* is distinctly smaller. All this would, no doubt, be ascribed by the youthful hearers to the fancies of old age, as nothing could be more certain, in their experience, than that all the stars regularly revolved around the zenith in concentric circles free from any variation.

We usually think of the North and South Poles as being the coldest areas on the surface of the Earth. There is reason, however, to believe that this view is quite erroneous. Professor Silvanus P. Thompson, of

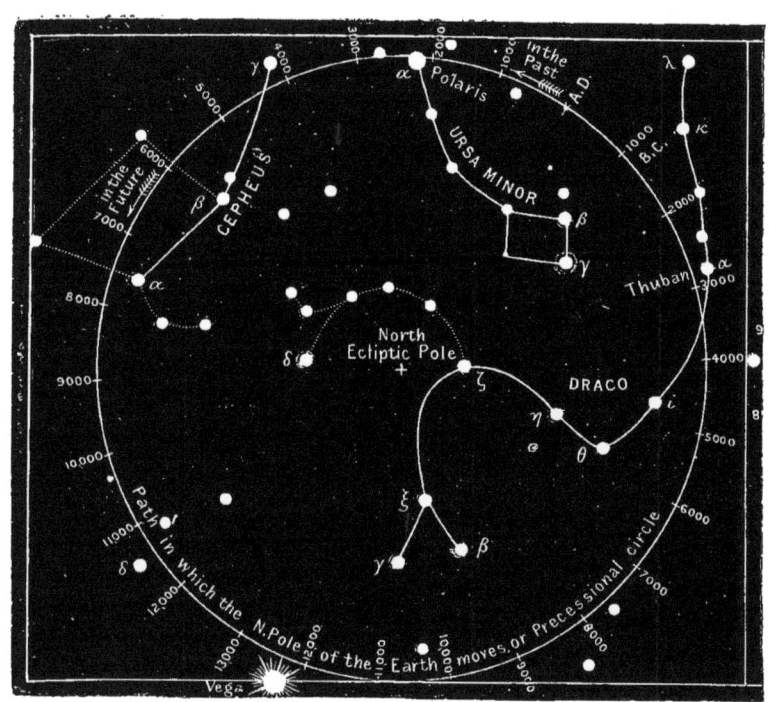

Diagram showing the path traced out by the northern extremity of the Earth's axis owing to the Precession of the Equinoxes.—p. 47.

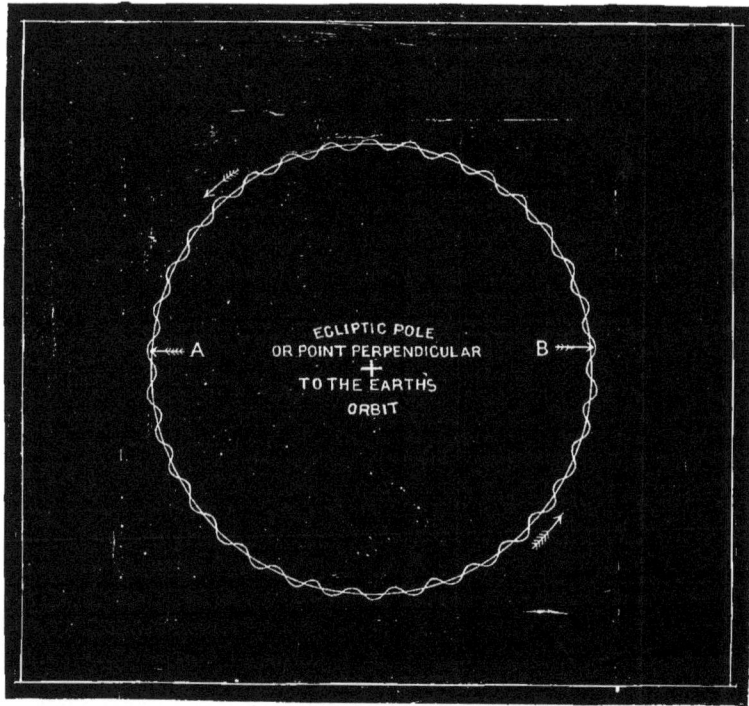

Diagram showing the path traced out by the Poles owing to nutation.—p. 48.

London Technical College, in writing regarding the polar aurora, says "the region of greatest cold in the Northern Hemisphere is known not to be at the geographical North Pole, but to be situated much more near to the North Magnetic Pole, which is also near the centre of the auroral belt or ring." It has been found that lines drawn on a map of the polar regions connecting places having the same mean annual temperature—the lines known as "isotherms"—show the existence of two poles of cold in the Northern Hemisphere, one in North America and the other in Siberia. There is reason to suppose that these are approximately the positions of the North Magnetic Pole and of what is known as the Asiatic Magnetic Focus. Each position is considerably removed from the North Geographical Pole, the North Magnetic Pole being in about latitude 70° 5′ N., and longitude 96° 43′ W., and the Asiatic Focus being in Northern Siberia. The lowest temperature ever recorded on the Earth's surface occurred at Verkhoyansk, in Northern Siberia, which is accepted as being in the region of greatest cold of, at any rate, the Old World. It is a penal settlement, with a population of about 350, and is in latitude 67° 34′ N., and longitude 134° 20′ E. In January 1883 the thermometer fell as low as 90° below zero F., which is 28° lower than the minimum temperature experienced by Dr. Nansen in his polar expedition. The average temperature of the three winter months at Verkhoyansk is said to be 53·1° below zero. The conjectured position of the Asiatic Magnetic Focus is not very greatly removed from the geographical situation of this inhospitable settlement.

As the North Pole, therefore, is greatly removed from

the positions of greatest cold, it seems probable that there the mean temperature will be appreciably warmer than it is in these positions. During the six months of exposure to the solar rays, it cannot be doubted that the temperature at the North Pole will be at least comfortably warm. Certainly the Sun never reaches a very high position in the heavens, but yet its greatest elevation, at the June solstice, is as great as it is in London in the beginning of November, or as great as it is in Florence at mid-winter. The mean summer elevation of the Sun at the North Pole is about 11° 45′ above the horizon, which is about the elevation it reaches at Sunderland, in England, at the winter solstice. The mean summer elevation of the Sun at the North Pole is thus considerable, and we have to keep in mind that, for six months continuously, day and night, the Sun remains in the heavens pouring light and heat on the Pole.

In other parts of the polar regions there is not the same continuous exposure to the Sun. The period of continuous exposure lessens with the separation from the North Pole, until, at the Arctic circle—at about 23½ degrees from the Pole—it is only for a few days, at midsummer, that the Sun does not set. Thus at the North Pole, the nightly radiation of heat, which occurs in those parts of the polar area where the Sun sets for a part of the twenty-four hours, will not occur. During the whole duration of the six months of sunshine, the Pole will be receiving more or less heat, and the surface heated will not be cooled periodically by recurring intervals of absence of the Sun. There is thus reason to suppose that, during the polar summer, the North Pole will be by no means a region of excessive cold, but that, on the contrary, it will have, with an excess of sunshine, a considerable degree of heat.

The contrast between the continuous sunshine of the summer six months, and the absence of the Sun for the winter six months must be terrible. Even in the high temperate latitudes, the lengthening of the noon-day shadow, as the Sun passes southward, is viewed with regret, although, in these latitudes, the Sun, when farthest south, appears for some time daily above the horizon. What then must be the feelings of the inhabitants of the polar regions as they see the Sun sink lower and lower in the heavens, until at last it disappears for the winter?   At the North Pole alone, in the Northern Hemisphere, does it sink to rise no more until the following season.   In other polar latitudes its winter disappearance is preceded by daily shortening periods of reappearance.   At the North Pole the Sun rises only once and sets only once during the whole year.

When the Sun at last sinks below the horizon, there must be a great fall of temperature at the North Pole. As the long winter night continues, the heat received during the summer will be lost by radiation, and the surface will gradually become chilled. It is known, however, that a great expanse of ocean exists in the polar regions, whether or not it actually covers the position of the Pole itself. The sea is more equable in temperature than the land, not being so receptive of solar heat, and not being so quickly cooled by radiation. From this it may follow that the variation of temperature at the Pole will not be so extreme as might be supposed. Of course, also, ocean currents may have a moderating effect. Thus, in our own neighbourhood, the Shetland

Islands have the advantage of the equable temperature of the sea, and of warm ocean currents, to such an extent that the difference between the mean summer temperature (about 53·5° F.) and the mean winter temperature (about 39·5° F.) is only about 14° F. Although they are situated mainly to the north of the 60th parallel, and thus only about six degrees south of the Arctic circle, their mean winter temperature is about half a degree higher than that of London, which is about 8½ degrees of latitude farther south. There may thus, also, be mitigating circumstances at the North Pole as regards the rigour of the winter temperature.

It has been conjectured that the waters of the Gulf Stream, which in their north-eastward course have such an effect in raising the mean winter temperature of the climate of Shetland—which, according to latitude, should be about 3° F.—continue in a northward course after they pass the Shetland Islands. They are supposed to sink below the surface, through their excess of salt, as they encounter the ice-laden waters from the polar regions, and it is thought that they may rise to the surface in the neighbourhood of the Pole itself, while they still retain a portion of the heat borne along from the Gulf of Mexico. Dr. Nansen refers to the Gulf Stream as being one of the chief currents running into the Arctic Ocean. He states that it passes north-eastward over the Shetland-Faröe-Iceland ridge, along the west coast of Norway. It divides into separate branches before it passes into the North Polar Basin, where it flows gradually northward and eastward (on account of the rotation of the earth), below the cold but lighter layer, 100 fathoms thick, of polar water. It fills the whole Polar

Basin, from a depth of 100 or 120 fathoms to the bottom, with Atlantic water.   Whether the water of the Gulf Stream rises to the surface again in the neighbourhood of the Pole can only be conjectured.   In any case the difference between the summer and winter temperature at the Pole must be extreme, although it is quite conceivable that the ocean at, or near, the Pole may not be so completely given over to ice as in the polar regions of lower latitudes.

Of course, temperature is generally more or less dependent on the direction of the winds.   This brings us to one of the outstanding peculiarities of the North Pole.   The only possible direction from which the wind can blow at the Pole is due south.   The compass declination relates solely to direction on the surface of the Earth, and thus, at the most northerly point on the Earth's surface, every direction is southwards.   Just as every road in Britain is said to lead to London, every point of the compass at the North Pole will point due south.

It is manifest that this curious state of matters is more theoretical than practical.   Although, geographically, every direction would be southwards, a person at the North Pole would be fully conscious of the existence of the four cardinal directions, which we are accustomed to distinguish as North, South, East, and West.   He would, however, have to find other names for them, if he wished to get over geographical difficulties.   He might, for instance, agree with the Astronomer Royal that the Sun when on the meridian of Greenwich is due south.   If so, when the Sun was on the 180th meridian, he would have to consider it as being due " opposite-south " or "anti-south," as it could not possibly be said to be in the north. A similar descriptive term might be found to indicate the direction of what we call the 90th meridian west longitude, and the 90th meridian east longitude.

The wind, therefore, although its direction would always be geographically from the south, might blow from the direction of either the meridian of Greenwich, or the 180th meridian, or the 90th meridian, west or east, or any intervening meridian.

The direction of the winds is greatly affected by the rotation of the Earth. When the atmospheric movement is from a region of high axial velocity, such as the equatorial region, towards a region of less axial velocity, the atmospheric current, having acquired the velocity of the region it is leaving, speeds eastward more rapidly than the adjoining part of the surface. It is therefore felt as a north-easterly or south-easterly wind, according as its movement in latitude is southward or northward. Conversely, an atmospheric movement towards a region of high axial velocity is felt as a westerly wind, with a northward or southward direction, according to its latitudinal movement.

Now, at the North Pole there is no axial velocity, as there is no displacement through the earth's rotation. The Poles are the only parts of the Earth at which this is the case. Any atmospheric movement to the Poles, from whatever part it may come, must thus necessarily come from a region of greater axial velocity, as well as from a southerly direction. It would, therefore, be of the nature which in other parts of the Earth would induce a south-westerly wind. At the Pole, however, different currents, all of the character which should induce what is known to us as a south-westerly wind, may come flowing in from all directions. It is difficult to form an opinion as to the result. As Sir Clements R. Markham, a former President of the Royal Geographical Society, says, in regard to the polar winds, "no general remarks upon them can be usefully made until our knowledge of the polar area is more complete." Undoubtedly, when the North Pole comes to be discovered, the observation of the meteorological phenomena will be of great interest.

At the Poles the force of gravitation is appreciably different from what it is at the equator. This arises from the peculiar shape of the Earth, owing to the polar flattening and the different local effect of the Earth's diurnal rotation. The force of gravitation varies according to the distance from the centre of the Earth and the local rotatory velocity. Through the polar flattening the distance of the Poles from the centre is less than the distance of any other part of the Earth's surface, and at the Poles there is no change of position through the Earth's rotation—and, consequently no relative centrifugal force,—while in their neighbourhood the change of position is very slight. Thus, at either Pole, a body falling to the Earth would fall more rapidly than at any other part. At the equator the increase of velocity gained by a falling body is 32·091 feet per second, while at the Pole it is 32·255. A falling body would, at the North Pole, have acquired a velocity of 32·255 feet in one second; a velocity of 64·51 feet in two seconds, and so on—the difference in the force of impact being therefore materially different at the Pole from what it is at the equator. At the equator the force of gravity is least, while at the Pole it is greatest; and, between these positions, it gradually increases with the distance from the equator and the approach towards the Pole, the velocity acquired by a falling body in one second being at Greenwich 32·191 feet per second. This does not mean that a body falling from a state of rest will fall the distance mentioned in the first second. It will, in fact, only fall half the distance, but, at the end of the first second, the *velocity acquired* is as stated.

From the same cause it arises that a pendulum, in order to beat seconds, has to be of greater length at the Pole than in any other latitude. It is shortest at the equator, and gradually lengthens with approach to the Pole. At the equator, the length of a pendulum vibrating in one second is 3·2514 feet, while, at the North Pole it is 3·2682 feet, a difference of ·0168 of a foot, or rather more than one-fifth of an inch. The required length is decided by the force of gravity, and the centrifugal force arising from the Earth's rotation. If a pendulum were made to oscillate at the North Pole, the terrestrial meridians would necessarily pass under it, and it would appear to describe a complete circle in the period of the Earth's diurnal rotation.

It necessarily follows that, at the North Pole, a body will weigh appreciably more than at the equator. If we divide the increase of velocity gained by a falling body at the Pole—32·255 feet per second—by the corresponding increase at the equator—32·091 feet per second—we shall get the weight of a body at the Pole in comparison with its weight at the equator, taking the latter as 1. Thus a body which, at the equator, weighs 1 lb., would, at the North Pole weigh 1·005 lbs. Consequently, a person whose weight at the equator is, say, 14 stone—that is, 196 lbs.—would, at the North Pole, weigh rather more than 14 stone 1 lb.—that is 197 lbs. As, however, the weights used in a scale would also weigh a correspondingly increased amount at the Pole, there would be no advantage in *trading* between the equator and the Pole, with the view of benefiting by the difference in weight on the exchange of commodities; unless, indeed, a spring-balance were used to measure the weight.

In all human affairs, we are accustomed to reckon the course of time by the Sun. The mean solar day may be said to be the chief unit of time-progression. "As sure as the rising of to-morrow's Sun" is accepted as an indisputable guarantee. It is difficult for us to realize that in this little planet, at no great distance from ourselves, there is an extensive area where these words have, for months at a time, absolutely no meaning, or a meaning directly the opposite of that attached to them by ourselves. At the North Pole, the solar day can have no existence, as the time-unit of the year. On such a basis, the year would consist of one day, as, during the year, there is only one sunrise and one sunset. The "solar day" at the North Pole would coincide with the year.

We, also, in our time-reckoning, fix the hour of noon, in the solar day, by the time of the Sun crossing our meridian, and the remaining hours are estimated therefrom, and are distinguished (in countries where the twelve-hours system applies) as either "ante-meridian" or "post-meridian." At the North Pole, however, every meridian meets, so that for six months the Sun is continuously crossing a meridian of the place. According to our system of reckoning time, it may be said that at the North Pole it is every hour of the day at one and the same time.

The difficulty in connection with the deciding of the hour of the day would, no doubt, be found by any person who was actually at the Pole, to be, like the difficulty as to the cardinal points, more theoretical than practical.

At the Pole, however, the mean time of the daily circuit of the heavens by the Sun is actually somewhat different from the mean time of its circuit in lower latitudes. This arises from the fact that, at the Pole, the mean time of the Sun's circuit would have to be reckoned on a six months' period—that is to say from the vernal to the autumnal equinox. It happens that the time when the Earth, in its orbital course, is most distant from the Sun occurs almost at the middle of this period—being in the beginning of July. As the orbital velocity of the Earth is least when the Earth is most distant from the Sun, the actual length of the solar day —being influenced by the arc of the orbital course travelled over by the Earth between successive returns of the Sun to the same meridian—is shortened appreciably below the average when the Earth is farthest from the Sun. From this it results that, at the North Pole, the mean length of the solar day, estimating on the six months' period when the Sun is visible at the Pole, is about 4·74 seconds less than the mean solar day of the lower latitudes, which is exactly 24 hours.

It is manifest that, at the North Pole, the course of the Sun would furnish a most inefficient time-reckoner, as the Sun is absent for about half the year, and the mean time of its circuit, during the remaining period, does not exactly conform to the mean time of other latitudes.

Evidently, if the North Pole were the abode of intelligent life, the course of time would necessarily be based on a more reliable celestial movement than the Sun would afford.   It would be based on a movement subject to no variation, so far as human experience has shown.   In brief, the time-keeper at the North Pole would be the rotation of the Earth itself, as shown by the apparent revolution of the stars, thereby produced—a time-keeper of absolute reliability.   No place on the surface of the Earth is better situated than the North Pole to calculate time by the true period of the Earth's rotation—the sidereal day.   In fact, during the six months of winter, the stars are a clock placed in the heavens, a glance at which will reveal the time.   The centre of this clock is the zenith, and any pair of stars fairly in line from the direction of the zenith towards the horizon may be considered as the hand of the clock.   Such a pair of stars we have in the *Pointers* of the Plough, or in *Pollux* and *Procyon*.   In fact the position of any prominent star—such as *Betelgeux* in Orion—which revolves a short distance above the horizon, would itself be sufficient to show what we may call " the time of day."

The reckoning of time by the apparent revolution of the stars would, of course, result in a varying divergence from solar time.   The sidereal day is about 3 minutes 55·9 seconds shorter than the mean solar day, so that this difference would occur in each day's reckoning. Consequently in the year, of about 365¼ solar days, the difference would amount to exactly one additional sidereal day.   Thus, in the reckoning of the hours by sidereal time, which would be the only practical method at the North Pole, the year would consist of 366 days, being the number allowed in the lower latitudes to leap years only.

We have noticed that the mean time of the Sun's revolution of the heavens in the polar summer is about 4·74 seconds shorter than the length of the mean solar day on the yearly average, being therefore about 23 hours 59 minutes 55·56 seconds. This is 3 minutes 51·16 seconds longer than the sidereal day, so that, during the polar summer, when the stars are hidden by the light of the Sun, it would be necessary to subtract 27 minutes every week from the hour as indicated by the place of the Sun, in order to preserve the sidereal time-reckoning.

Dr. Alexander Buchan mentions that there are on the surface of the Earth two regions of high atmospheric pressure—or excess of air—the one north, and the other south of the equator, passing completely round the globe as broad belts; and that there are three regions of low pressure—or deficiency of air—one round each Pole, and the third at the equator between the two belts of high pressure. As is well known, the Earth is not a perfect sphere, but is flattened at the Poles, the polar diameter being about 7,900 miles, while the equatorial diameter is about 7,927 miles. The polar flattening and equatorial bulge are believed to have been occasioned by the rotation of the Earth while it was still a molten mass. It seems most probable that the low atmospheric pressure around the Poles and the belts of high pressure at the equator, are also occasioned by the rotation of the Earth: that the same causes which produced the polar flattening on the molten earth still operate to produce a polar flattening of the gaseous envelope which forms the atmospheric covering of the Earth.

It is, of course, impossible to tell whether or not there is land at the North Pole. Sir Clements Markham, in commenting on the memorable drift of Dr. Nansen's ship, the *Fram*, in the polar ice,—a drift which was continued during three winters (1893-6)—writes as follows:—

" Nansen's chief discovery is that there is a very deep ocean to the north of the Franz Josef Group, continuous with that to the north of Spitzbergen, and that this deep sea has a relatively warm temperature in its depths. The ice-bearing ocean extends at least as far as the Pole, and the time occupied by the ice in drifting across it from the Siberian side to Spitzbergen is about three years. The existing evidence points to the conclusion that there is a continuous drift from the Eastern to the Western Hemisphere, across an ice-covered ocean uninterrupted by land of any magnitude."

As illustrative of the comparative warmth of the water in the polar winter it is said that men who have happened to fall into the sea within the Arctic Circle have found it at first rather warm and agreeable by contrast with the prevailing temperature of the atmosphere. They have no sooner scrambled out again than the keen frost has caused their garments to become as hard as boards.

The whole fringe of the European, Asiatic, and American coast-lines within the Arctic Circle has been found to be inhabited, and the inhabitants have spread up the shores of Boothia, and up both sides of Davis Strait and Baffin's Bay. Spitzbergen, Franz Josef Land, and Nova Zembla are uninhabited, except that occasional summer visits are made to the southern shores of the latter group of islands. It is said that the most northern people in the world are a tribe to whom Sir John Ross,

the celebrated explorer, gave, in 1818, the name of
" Arctic Highlanders." Their stations range along the
Greenland coast from 76° to 79° N. These northern
peoples live mainly on sea animals, and rarely wander
from the coast. Recent exploration does not appear to
have revealed any trace of inhabitants at higher latitudes
than those mentioned.

The Arctic Regions are by no means destitute of
animal and plant life, even in the highest latitudes
reached. Trees, however, in the proper sense, do not
occur within the Arctic Circle—the Arctic species of
willow and birch being low-growing plants, herbs, indeed, rather than shrubs. The development of organic
life is, as might be expected, comparatively poor in those
parts of the Artic Ocean which are continuously covered
by ice. It has been found that the bottom deposits contain unusually little carbonate of lime of organic origin.
Dr. Nansen mentions that there is " exceptionally little
plant life in the sea under the ice covering, and the
animal life both near the surface and in deeper strata is
very poor in individuals, whilst it is comparatively rich
in species." Near the outskirts of the Arctic Ocean,
where the sea is more open, marine plant life and
animal life are unusually abundant. Nansen suggests
that the North Polar Basin may be considered as a lung
or reservoir in ocean circulation, where the water produces little life, but itself becomes re-invigorated, and
he points out that the greatest fisheries in the world
occur in the places where the waters of the Arctic and
southern seas meet. The outskirts of the Arctic Ocean
are exceptionally rich in mammals and birds, but, in the
interior, such life is far from being abundant. There is,

however, no reason to suppose that the locality of the North Pole will be destitute of life. There is, indeed, some reason to think that such life may be rather more abundant than polar exploration would suggest, in the yet unexplored regions of the polar area, just as there is ground for believing that the intensity of the cold of the polar winter will there not be so extreme as it is in some of the latitudes which have been the scenes of recent exploration.

The axis of the Earth, is, of course, a merely imaginary line, about which the Earth rotates. The North Pole might, therefore, naturally be assumed to be a mere point on the surface, as being the northern extremity of this imaginary line. It has, however, been found that the Earth's axis is not perfectly stable, but oscillates periodically to the extent of about ·3 of a second of arc. Now if we take the length of the degree of latitude as representing 69 miles—and in reality it is, at the Pole, nearly 69½ miles—we find that this distance—·3 of a second of arc—is equivalent to 30·36 feet. It would, therefore, appear that the North Pole oscillates periodically in, and around, a distance on the Earth's surface, having a diameter of about 30·36 feet.

Although the North Pole, as the terrestrial termination of a merely imaginary line is, no doubt, at any particular moment a mere point on the Earth's surface, this movement of oscillation affords some warrant for asserting that, considered geographically, the Pole has a diameter of 30·36 feet; and in view of the relation between the diameter and the circúmference of a circle, we may say that the North Pole measures in circumference about 95·38 feet. This does not greatly exceed the girth of some of the giant redwood trees of California. The bark of one of the finest of these trees, which was known as "the mother of the forest," is in the Crystal Palace. This tree measured at the base 90 feet in circumference. Thus, what we may describe as the "circumference" of the terrestrial Poles — or the space through which they oscillate—only slightly exceeds the circumference of one of these mammoth trees.

It is not inconceivable that a time may come when the difficulties of access to the North Pole will be surmounted, say, for instance, by the solution of the problem of ærial navigation. Our descendants may then be able to indulge in the luxury of a summer trip to the Pole. There is certainly some reason to suppose that, in the height of summer, the climate at the North Pole is pleasant, warm, and bracing, and that the Pole would, therefore, if convenient of access, and if on a land surface, be a desirable summer retreat.

When we consider the great progress of geographical discovery in late years, and the strenuous efforts, and the international rivalry which continue in relation to the discovery of the North Pole, the conclusion seems almost irresistible that, before long, the actual discovery of the Pole will be added to the achievements of the daring army of explorers. The discovery of the North-West Passage was for generations vainly attempted, but it yielded up its secret to Sir Robert M'Clure in 1850-51. The discovery of the North-East Passage was, in like manner, the object of prolonged exertion, but persistence again prevailed and this geographical puzzle was solved in 1879 through the efforts of Baron Nordenskjold, a native of Finland. There are, indeed, comparatively few geographical areas now remaining unexplored, and, as the number lessens, the efforts of the undaunted band of pioneers become more concentrated. All the civilized nations of the world are now interested in the discovery of the North Pole, and the fortunate discoverer will certainly gain a well-merited position in the list of renowned explorers. The British have taken a foremost place in geographical pursuit in the past, and, in Arctic exploration, they have been specially prominent. Whether they shall be the first to arrive at the winning post of the northern polar regions, time alone can show. In any case there will be no international jealousy of the hero who shall first penetrate the great ice barrier, and in name of his country take possession of the crown of the Northern Hemisphere.

THE APPARENT ENLARGEMENT OF HEAVENLY BODIES NEAR THE HORIZON.

F

SYNOPSIS.

Apparent enlargement more generally observed in the case of the Moon than of the Sun—Actual distance of Sun and Moon when on the horizon as compared with their distance when at the zenith—Variation in distance of Sun and Moon conducive to enlargement at zenith rather than at horizon—Antiquity of the problem—Explanations put forward in the past—Recent discussion in *Astronomical Journal* —Four different explanations suggested—Apparent objections to certain views—Whether the phenomenon may be due to atmospheric causes—Estimates of height of atmosphere—Increased depth of atmosphere with approach to the horizon in our observation of the Sun and Moon respectively—Increased density and humidity of atmosphere in horizontal view—Geographical bearings on the extent of atmosphere in different lines of sight—Changes in relation to atmosphere according to line of sight explanatory of phenomenon.

THE APPARENT ENLARGEMENT OF HEAVENLY BODIES NEAR THE HORIZON.

Now and then most people have noticed that the Moon when near the horizon appears to be decidedly larger than when it is at a considerable altitude. At times, indeed, the evident enlargement almost compels attention. The Moon, at the horizon, occasionally seems to the eye to be two or three times as large as its apparent size when high in the heavens.

Although the popular observance of this apparent enlargement more commonly occurs in relation to the Moon than in relation to the Sun, it would seem to be the case that the Sun is, in this respect, in exactly the same category as the Moon, and thus is sometimes seen, when in the neighbourhood of the horizon, very noticeably larger than when high in the heavens. Probably the explanation of the enlargement being more generally observed in the case of the Moon arises from the Moon being more easily looked at than the Sun, even when the latter is low down in the sky and partially shaded.

As the actual distance of either the Sun or the Moon may be taken generally as being somewhat *less* when on the meridian of the observer than when on his horizon, an opposite result might naturally be expected;

that is to say, any apparent enlargement might reasonably be looked for high in the heavens as compared with the horizon, rather than at the horizon as compared with high in the heavens. If we disregard the varying distance of the Sun and Moon, and consider the distance separating them from the Earth as constant, then, evidently, each would be nearest to any particular place when crossing the meridian of that place. Although the distance of the Sun and the Moon from the Earth is continuously changing through the orbital movements of the Earth and the Moon respectively, this change cannot appreciably affect their apparent size in the short interval which elapses between the times of their crossing the meridian and reaching the horizon.

The actual variation in the distance of the Moon from the centre of the Earth, during a lunar revolution, is about 31,355 miles, being the difference between the maximum distance—about 252,948 miles—and the minimum distance—about 221,593 miles. The Moon takes half the time of its revolution—being about 13 days, 15 hours, 52 minutes—to pass from its maximum to its minimum distance. Thus, when the Moon's distance is lessening, the reduction in distance during the mean time of the Moon passing from the meridian to the horizon—say about 6½ hours—would be about 622 miles. By the daily rotation of the Earth, however, we *increase* our local distance from the Moon, in the interval which elapses between its appearance on the meridian and its appearance on the horizon, by nearly 4,000 miles. This arises simply from the fact that we, by the rotation of the Earth, are turning away from the

Moon. This result of the Earth's rotation may readily be tested by trigonometrical methods.* If we take the mean distance of the Moon from the centre of the Earth as 238,818 miles, and the semi-diameter of the Earth as about 3,963 miles, we get the local increase in the distance of the Moon, which results from its position being changed from the zenith to the horizon, as about 3,930 miles. The Moon's orbital progression when its distance is actually decreasing, will, during the interval which elapses between its appearance on the meridian and on the horizon of any particular place, lessen its separation from the centre of the Earth by about 622 miles, but the Earth's rotation in the same time increases the Moon's distance from *the place of observation* by about 3,930 miles, so that, notwithstanding the Moon's movement towards the Earth, it will be about 3,308 miles farther away from the place of observation when it reaches the horizon than when it crossed the meridian. This is, of course, an extreme illustration, as we have assumed that no change occurs in the declination of the Moon, which could not be the case, and have started with the Moon in the zenith which, in the higher latitudes, does not occur. The result arrived at, as regards the amount of increase of distance, is therefore exceptional, but the conclusion can be excepted to the extent of proving that the increased distance produced by the rotational movement of the Earth is invariably such as to far more than counterbalance any reduction in distance resulting from the lunar movement during the time the Moon takes to pass from the meridian to the horizon.

* See page 72.

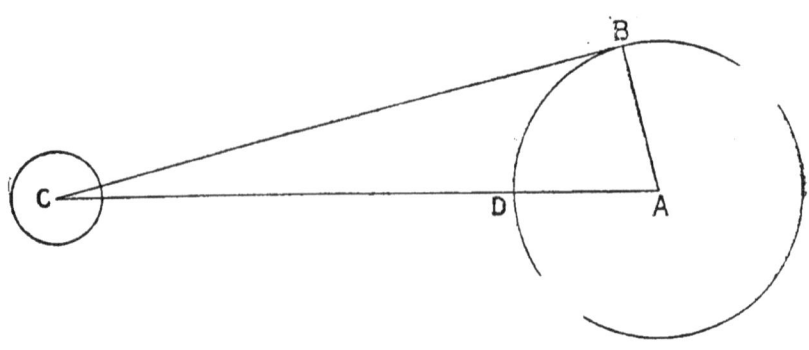

Diagram showing the local change in the distance of the Moon resulting from the Earth's rotation.

Larger Circle =The terrestial equator.
Smaller Circle =The Moon.
A =The centre of the Earth.
C =The centre of the Moon.
B =Position at equator having the Moon on the horizon.
D =   „   „   „   „   in the zenith.
AC =238,818 miles, being the Moon's mean distance from the centre of the Earth.
AB and AD are semi-diameters of the Earth at the equator, and are each 3,963 miles in length ; the angle at B is a right angle.

$BC^2 = AC^2 - AB^2 = 238,818^2 - 3,963^2$
$= 57034037124 - 15705369 = 57018331755.$

$\therefore BC = 238,785$ miles.

The Moon's mean distance from the surface at the equator, when in the zenith, is 234,855 miles, being its mean distance from the Earth's centre (238,818 miles) less the equatorial semi-diameter of the Earth (3,963 miles) ; while its mean distance from the surface at the equator, when on the horizon, is 238,785 miles, so that the rotation of the Earth, in changing the place of the Moon from the zenith to the horizon increases its distance by 3,930 miles.

It thus appears that, even when the Moon in its orbital course is *decreasing* its distance from the Earth, it is, nevertheless, considerably nearer any place of observation when it is crossing the meridian of that place, than it is a few hours later when it reaches the horizon.

In the case of the Sun, a somewhat different state of matters occurs. . The annual variation in the distance of the Sun is about 3,100,000 miles, and the time which intervenes between the maximum and the minimum distance

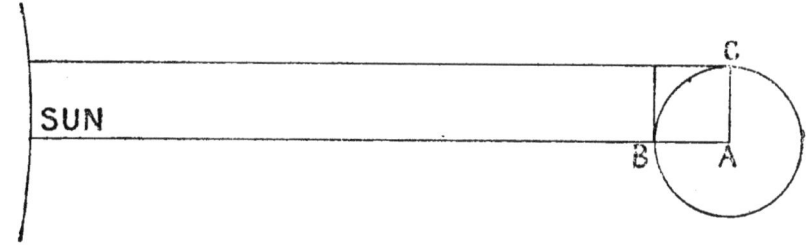

Diagram showing the local change in the distance from the Sun resulting from the Earth's rotation.

Circle = equator.
A = centre of Earth.
B = position having Sun on meridian.
C = position having Sun on horizon.

It is evident that at C the distance of the Sun is greater by the length of the terrestrial radius (about 3,963 miles) than it is at B.

—that is from, say, 4th July to 1st Jannary—is about 181 days. Thus the mean *daily* variation in the Sun's distance during that period is about 17,127 miles, and the variation in the time which elapses between the Sun crossing the meridian and reaching the horizon—say six hours—would consequently be about 4,282 miles. The local increase in the distance from the Sun, would, in the same interval, be about 3,963 miles. This arises entirely from the rotation of the Earth, but the lines of sight from positions having the Sun in the zenith and on the horizon respectively, do not, as in the case of the Moon, form any appreciable angle. Owing to the immense distance of the Sun (about 92,897,000 miles on the mean) the angle made by two lines drawn from different parts of the Earth to the Sun is almost inappreciable. It is evident, however, that—at any rate at the equator—the rotational movement from the position which has the Sun on the meridian to the position which has the Sun on the horizon, increases the *local* distance from the Sun by one half of the diameter of the Earth; being by about 3,963 miles. Thus, the Sun on reaching the horizon would, if its distance were decreasing, be in reality somewhat nearer an observer than when, six hours earlier, it crossed his meridian. The reduction in its distance would, however, be only about 319 miles, being the difference between the mean reduction in the distance resulting from the Earth's orbital movement at the time of lessening separation (4,282 miles) and the increased separation at the place of observation resulting from the Earth's axial movement (3,963 miles.) This nett reduction of 319 miles, being about 1-291,000th of the mean distance, may be considered as absolutely negligible in so far as it could affect the visible appearance of the Sun. It is indeed much the same as if an object at a distance of a mile approached nearer to the observer to the extent of about one-fifth of an inch. Any variations arising from geographical position are necessarily very insignificant, and may be disregarded. We may, therefore, say that when the distance of the Sun is decreasing the Sun is, to all intents, at the same distance from any place of observation when it is crossing the meridian, as it is six hours later, when on the horizon.

Thus, we find, in the case of the Moon, that, even when

its distance from the Earth is on the decrease, it is considerably nearer a place of observation when crossing the meridian than it is when it reaches the horizon; and, in the case of the Sun, that, when the distance separating it from the Earth is decreasing, it is virtually at the same distance from a place of observation when crossing the meridian, as it is when it reaches the horizon six hours afterwards, the reduction in distance resulting from the orbital movement of the Earth being, as regards any place of observation, almost entirely cancelled by the axial movement away from the Sun.

If this then is the case when the relative distance of separation is on the *decrease*, it is clear that, when the distance is on the *increase*, both the Sun and the Moon must be considerably nearer any place of observation when they respectively are crossing the meridian of such a place, than when they next descend to the horizon.

All these facts go to show that we might naturally expect any apparent enlargement of either the Sun or the Moon to occur on, and in the neighbourhood of, the meridian, and any apparent diminution in size to occur on, and in the neighbourhood of, the horizon. In the case of the Sun, indeed, when its distance is lessening, we might suppose that no apparent enlargement or diminution in size could occur in either position as the one set of circumstances balances the other. On the other hand, when the distance is increasing, the Sun should clearly fall into the same category as the Moon.

Evident as this reasoning appears to be, daily experience proves that it cannot be accepted. The Moon, when on the horizon, notwithstanding its increased distance, appears considerably larger than when on the meridian.   The Sun, when on the horizon,. notwithstanding the varying effects of its change of position, also appears to be considerably larger than when it is at its highest point in the heavens.

Since the time of Ptolemy, the celebrated astronomer, who flourished in the second century of our era, explanations of various kinds have been put forward to account for these phenomena, which have thus received attention for nearly two milleniums.   Notwithstanding this, no general agreement as to the cause of the apparent enlargement at the horizon can yet be said to have been come to.   Let us now consider the different explanations, which have been suggested.

Descartes, the celebrated French philosopher, appears, early in the seventeenth century, to have first suggested the explanation, which, to this day, is the most generally accepted.   His idea is, that, when the Sun or the Moon is on the horizon, we unconsciously judge its size by terrestrial objects, and are thereby the victims of what is simply an optical illusion.   His views, and those of many eminent astronomers since his time, are summarized by Mr R. A. Proctor.

"The opinion which we form of the magnitude of a distant body does not depend exclusively on the visual angle under which it appears, but also on its distance ; and we judge of the distance by a comparison with other bodies.   When the Moon is near the zenith there is no interposing object with which we can compare her, the matter of the atmosphere being scarcely visible.   Deceived by the absence of intermediate objects, we suppose her to be very near.   On the other hand, we are used to observe a large extent of land lying between us and objects near the horizon, at the extremity of which the sky begins to appear; we therefore suppose the sky, with all the objects which are visible in it, to be at a great distance. The illusion is also greatly aided by the comparative feebleness of the light of the Moon in the horizon, which renders us in a manner sensible of the interposition of the atmosphere. Hence the Moon, though seen under nearly the same angle, alternately appears very large and very small."

An interesting correspondence on this subject, initiated by a paper by Dr. Cecil G. Dolmage, F.R.A.S., appears in the *Journal of the British Astronomical Association* for 1899–1900. Dr. Dolmage mentions that the explanation generally accepted is, as we have noticed, that we estimate the size of celestial bodies by comparing them with terrestrial objects in the same field of view, that such comparison is naturally most simple when the celestial bodies are at, or near, the horizon, and that, as we know that the terrestrial objects in view are in reality much larger than they appear, we unconsciously exaggerate the size of the celestial bodies by a species of sympathy.

Dr. Dolmage mentions that the apparent enlargement is not by any means confined to the cases of the Sun and Moon, and that it is very noticeable in the case of stars situated a little' distance apart. The distance of their separation is apparently much increased when they are in the neighbourhood of the horizon, from what it appears to be at the greater altitudes. He illustrates his remark by the relations between *Capella*, in the constellation Auriga, and the large star near it, in the same constellation, known as *Beta Aurigæ*. These two stars appear much farther apart when low down, than when high up in the sky. This apparent widening out is also, according to Dr. Dolmage, noticeable during the early evenings of

October in the setting of "that magnificent triangle of stars, of which *Arcturus* occupies the lowest angle." He expresses the opinion that the evident and increasing distortion of the relative parts of the triangle in its gradual descent to the horizon cannot but strike any observer familiar with its appearance when high up in the sky. At the greatest altitude the appearance resembles an equilateral triangle. In its descent to the horizon the triangle becomes more and more isosceles, being drawn downwards in the direction of *Arcturus*. At the same season of the year, a corresponding change is observable in the appearance of the constellations Perseus and Andromeda. As they mount upward towards the zenith a gradual closing together of the larger stars in each constellation appears to take place; as they descend towards the horizon the converse appears to occur.

Dr. Dolmage enquires what is the reason of these appearances, which are known to be illusory since micrometric measurements fail to explain the matter in any way. He suggests that the true explanation is not that generally accepted, of an unconscious exaggeration of the size of the celestial bodies by the observer as they approach the horizon, but that it is to be found in the apparent curvature of the celestial vault. We may give his suggestion in his own words :—

"The explanation which I desire to submit is as follows :— Let an observer look up towards the zenith, and consider the expanse of sky presented to his gaze. He will notice that it contains *practically no trace of curvature;* in fact, it will appear to him as flat as the roof of a room. This is, evidently, because his field of view is not extensive enough to include any appreciable portion of the dip of sky towards the horizon.

In fact had he lain upon his back all his life and known nothing at all about the horizon, he would never consider the expanse of sky above him as anything other than a plane. Next, let him lower his eyes gradually towards the horizon. In accordance as he does this he will notice that, on account of the dip of the sky in front, and, particularly, on either side of him, coming more and more into the field of view, the plane heaven which he has just been looking at will, little by little, assume the appearance of concavity, this appearance of concavity rapidly increasing, and finally reaching its maximum when the gaze is directed to a point on the horizon. Now, let us see what effect this change will have upon a definite portion of the sky as seen by an observer. It must be quite clear to anyone, who for a moment considers the matter, that the distance between markings upon a plane *elastic* surface will be increased in proportion as the plane is bodily hollowed out so as to assume a concave shape, *the boundaries of the plane being always preserved at the same initial distance from each other* . . . This then gives the clue to what takes place when an observer views the heavens. The eye looking zenithwards perforce considers the expanse as a plane : then, as the gaze is lowered, and the dip around the sky towards the horizon comes increasingly into view, the eye *unconsciously* considers this plane as gradually becoming more and more concave towards it ; therefore the total area of the *field of view* of sky (regarded chiefly in its lateral extent), and, in consequence, the area of any definite portion of it, is considered as having enlarged."

Dr Dolmage illustrates his suggestion by the supposition of a large sheet of indiarubber stretched tightly between two posts, having marks made upon it a short distance apart. If the sheet is hollowed out in the opposite direction from an observer, the distance between the marks, and the surface expanse of the sheet, will both be increased. Similarly, to an observer inside a balloon, any surface markings would appear to expand with the gradual inflation of the balloon. He supports his suggestion by his own experience. He observed the newly-risen Moon from a brilliantly lighted room, whereby the Earth, sky, and horizon, were all concealed from view. The Moon, under these circumstances, appeared to have no enlargement. When viewed immediately afterwards in the open air the accustomed enlargement was quite evident, and also the marked concavity of the sky, in which, according to Dr Dolmage, the explanation of the apparent enlargement is to be found.

The correspondence which followed Dr Dolmage's paper, shows how very unsettled scientific opinion on this subject still is, in spite of the great antiquity of the problem and the supposed very general acceptance of the views propounded by Descartes, more than 250 years ago.

One writer, Mr R. G. M. Browne, suggests that the illusion is, in all probability, occasioned "by the external form and internal structure of the instrument employed in beholding it; that is to say, the eye itself." He states that the eye-image received by the brain appears larger when the view is a horizontal one than when it is at a height above the horizon; and, proportionately with the increase of height, the size of the image transmitted through the eye to the brain seems to become diminished.

The suggestion put forward in the passage which we have quoted as summarizing the views of Descartes—that the horizon appears more distant than the sky overhead —is dissented from. Mr Edwin Holmes contends that, with a clear sky, the zenith distance appears much greater than the horizon distance. He put the question to one or two people, whether, with a blue sky, the sky looks more distant overhead or low down, and received the answer that "low down looked about as distant as the most distant terrestrial object, but that overhead seemed far more distant," an answer which agreed with his own conception.

Proctor, in the Encyclopædia Britannica, expresses the opinion that "the apparent enlargement of the Sun and Moon near the horizon is an optical illusion, connected in some measure with the atmosphere." This view finds strong support in the discussion to which we have referred. Mr James D. Hardy contributes a letter which refers to this belief. We may quote a portion of his communication :—

"Firstly, we must thoroughly understand under what conditions the Moon does appear enlarged when on the horizon, because it is not at all times so evident. This is mostly the case when, after a hot clear day, there is a considerable amount of evaporation and condensation at sunset. The enlargement is then most observable at sea, as we then see the Moon at rising (the full Moon is the more evident), through a greater depth of atmospheric conditions, and then there is nothing to affect its apparent size, or lead us astray, by comparison with other objects, and its size is affected more or less solely by the action of its rays on and through the heavily charged moist atmosphere of condensation. It is only repeating common knowledge, but it is advisable to make the statement, that, when rays of light pass through air laden with particles of moisture they are refracted and reflected . . . We do not see the Moon as it really is, but where the angles of reflexion make it appear to be, and the combined effect of these refracted and reflected rays is to increase the apparent size of the Moon."

Although Mr. Hardy mainly confines his argument to the Moon, it is of course applicable to all the celestial luminaries, in varying degrees.

At the present day therefore, although the investigation of the subject is of such great antiquity, no less than four different explanations are put forward, each of which has some scientific support. These explain the apparent enlargement as follows :—

(1) By unconscious comparison of celestial bodies when near the horizon with terrestrial objects in the field of view, whose size we are in a position to judge.

(2) By the apparently increased curvature of the celestial vault, as the horizon is approached.

(3) By the form and structure of the organ through which we receive the impression of enlargement—that is the human eye.

(4) By atmospheric causes.

It seems to be agreed that the extent of the apparent enlargement is of a variable character, that at times it is very noticeable, and at times either absent or scarcely perceptible. This appears to be a fatal objection to the second and third of the suggested explanations. The increased curvature of the heavens, as we approach the horizon, does not seem to be variable if the distance from the zenith is the same ; and the form and structure of the normal eye do not alter, in an irregular manner, from day to day. If, therefore, either of these explanations were correct and complete in itself, the enlargement perceptible should be constantly the same on each return of the Sun or Moon to a similar position on the horizon. Of course, however, either explanation might not unreasonably be accepted as contributing to the effect. In itself, however, each may be considered as quite insufficent to account for any variableness in the character of the enlargement.

The first explanation—that the phenomenon is occasioned by unconscious comparison with terrestrial objects—has received much acceptance, but it is evident, from the discussion to which we have referred, that many persons interested in the question have not accepted it. In fact, while it can by no means be disproved, it is somewhat unconvincing.

The fourth explanation—that the phenomenon is occasioned by atmospheric causes—appears on the whole to be the most convincing.

There can, of course, be no question as to the apparent enlargement being an optical illusion, as it is evident that the actual size of celestial bodies cannot possibly change, as they pass from the zenith to the horizon. It requires no argument to satisfy us that the enlargement is merely apparent.

The view that the apparent enlargement is occasioned by atmospheric causes received much support in the discussion in the *Journal of the British Astronomical Association.* Mr. Edwin Holmes writes that he believes it to be a fact that the Sun and Moon loom large on the horizon on many occasions with a hazy or smoky sky, but that he personally has never seen anything of the kind with a clear sky. He notes that neither photographs nor micrometric measurements made low down,

show any real enlargement, but he questions whether
these have been made with a misty atmosphere to produce enlargement. In his opinion, it is only under such
conditions that any apparent enlargement takes place.

Mr. John Turner, M.A., B.Sc., mentions that travellers
from this country to Switzerland commonly report a
sense of disappointment at the first view of the Alps.
The reason is that, owing to the clearness of the
atmosphere, the mountains appear much nearer than they
really are, and, the distance being under-estimated, the
size is also under-estimated. Thus a clear atmosphere
induces an under-estimation in size, while on the
other hand, objects loom large through a fog. In
Switzerland, as in a fog, our estimates will be corrected
by further experience, but a sudden change from one
condition to the other involves mistaken estimates. Mr.
Turner compares the zenith observations to views through
a clear atmosphere, and horizon observations to views
through a foggy atmosphere. In looking towards the
zenith, we are looking through a thin veil of atmosphere,
while in looking towards the horizon, we are looking
through a thick one. He states that the apparent
enlargement at the horizon is most pronounced when the
horizon is specially hazy : and that, on a very clear night,
the Moon at a high altitude looks smaller than it does
even at the same altitude when the atmosphere is less
transparent.

Let us now turn our attention for a little to the consideration of the atmosphere, and see what difference, in
relation to the atmosphere, occurs in our line of view,
according as a celestial body is observed at a high altitude
or near the horizon.

The atmosphere forms a complete covering to the Earth, and, doubtless, its outer bounds are fairly regular, and free from such hills and valleys as characterize the solid Earth. From this, it follows that the atmospheric covering of mountainous districts is thinner than the atmospheric covering of the neighbouring seas and valleys, the thickness of the atmosphere varying, indeed, in the same locality, according to the height of the land.

Notwithstanding the stupendous importance of the atmosphere, and the fact that in it we pass our existence, our information regarding its height is very vague. The late Professor E. Loomis, of Yale College, U.S.A., estimated the height of the atmosphere at about 500 miles. Dr. Alexander Buchan, of Edinburgh, a distinguished meteorologist, concluded that the height was from about 120 miles to 200 miles. Other writers estimate the height as low as from 40 to 50 miles. The actual height does not materially affect the subject we are now considering. We have to ascertain how the atmospheric envelope will intervene in our line of sight of heavenly bodies at their greatest altitude and at the horizon respectively; and what differences will be thereby occasioned in our view of these bodies in these respective positions.

We may now suppose that we are observing the Sun, or the Moon, from a position at the equator, and that the Sun, or the Moon, is in the zenith. It is clear that, under such circumstances, the thickness of the atmosphere intervening between us and the luminary, in our line of sight, is actually, so far as can be judged, the mean thickness of the atmospheric envelope which encircles the Earth, which, as we have seen, may be 500 miles, 200 miles, or less.

Let us now suppose the luminary to have sunk to the western horizon, but to be still directly over the equator. The Sun, or the Moon, will now be in the zenith about 90 degrees to the west of the position at which we have taken our stand. It is manifest that, in looking towards it, we are looking through a very much greater thickness of atmosphere than if it were directly overhead. We are, as it were, looking sideways through the atmosphere.

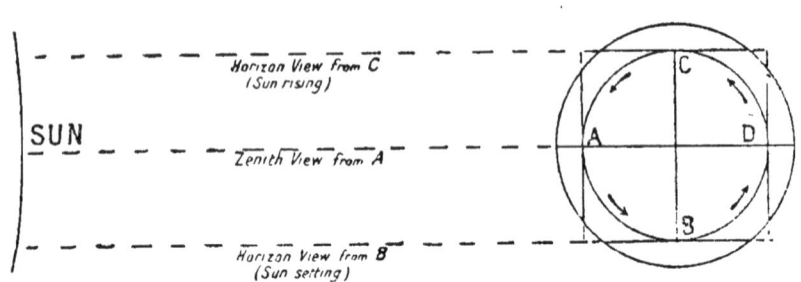

Inner circle = equator.
Outer circle = outer limits of atmosphere.

Diagram showing the increased extent of atmosphere in the line of sight, with approach to horizon.

This can be clearly and simply shown by a diagram. We can draw a circle representing the equator, surrounded by a slightly larger circle representing the outer limits of our atmosphere. The space between these two circles will then represent the atmosphere. If now, we draw two diameters in the inner circle at right angles to each other, we shall have, at the extremities of the diameters, positions on the equator separated by 90 degrees of longitude. We may call these four positions A, B, C, and D, respectively. If, now, we suppose the Sun, or the Moon, to be in the zenith at A, then, as B and C are each separated from A by 90 degrees of longitude, it will be on the horizon in each of them. In one case, it will be on the eastern, or morning, horizon; in the other case, on the western, or evening, horizon. As D is separated from A by 180 degrees of longitude, the zenith in D will be at the diametrically opposite part of the heavens from where it is in A. The zenith in A is the nadir in D. If it is noon in A, it is midnight in D. It is evident from the diagram that, in each position, when the line of sight is towards the zenith, it crosses the atmosphere at its smallest width in relation to the particular place of observation. When, however, it is directed towards the horizon, it passes through the greatest thickness of the atmosphere in relation to the place of observation. The extent of the atmosphere, in the line of sight between these two extremes, is proportioned to the separation from the zenith and the approach to the horizon. This applies equally to each of these positions.

For the sake of simplicity, we have illustrated our statement by the employment of positions situated on the equator. It would seem, however, that in this matter no distinction arises between positions on the equator and positions in other latitudes. The circle representing the equator may equally well be taken to be a circle of latitude at any distance from the equator, subject to the varying correction in connection with the position of the luminary on the horizon, occasioned by the changing seasons.

As it is therefore clear that, when we look towards the Sun or the Moon on the horizon, we are looking towards it through a much greater thickness of atmosphere than when we look towards it at the zenith, or at a fair altitude, we may now endeavour to obtain

some idea of the thickness of the atmosphere in our
line of sight towards the horizon, in comparison with
its thickness in our line of sight towards the zenith,
first dealing with the question in relation to our view
of the Sun.

Let us, as in our previous illustration, make a circle
representing the terrestrial equator, with a circle

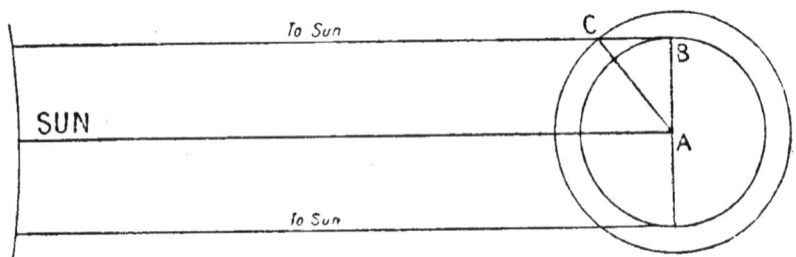

Diagram showing method of finding the comparative depth of the
atmosphere in the line of sight towards the horizon.

> Inner Circle = the equator.
> Outer circle = outer limits of atmosphere.
> Radius of Earth (AB) = 3,963 miles.
> BC = line of sight towards the horizon.

The height of the atmosphere is taken to be 160 miles (being the mean
of the estimates of Dr. A. Buchan).

$BC^2 = AC^2 - AB^2 = 4,123^2 - 3,963^2.$
$= 16999129 - 15705369 = 1293760.$
$\therefore BC = $ about 1,137 miles.

The corresponding figures, taking the height of the atmosphere to be
500 miles, 200 miles, or 50 miles, can be similarly worked out. They are, as
stated in the text, about 2,052 miles, 1,275 miles, and 631 miles respectively.

surrounding it, representing the limits of our atmosphere.
We can draw two lines towards the Sun, one from a
position having the Sun on the meridian and the other
from a position having the Sun on the horizon. Owing
to the vast distance of the Sun, these two lines will
make no perceptible angle, and must be considered

as parallel. Let us now draw a line from each of these positions to the centre of the Earth, and then draw a line from the centre of the earth to the point where the line from the position having the Sun on the horizon cuts through the outer limits of our atmosphere. This gives us a right-angled triangle—the right angle being at the position having the Sun on the horizon. As one of the sides of the triangle is one-half of the equatorial diameter of the earth, and is therefore about 3,963 miles in length, and another side is of the same length *plus* the height of the atmosphere, we shall, by *assuming* the height of the atmosphere, have the length of two sides of our right-angled triangle, and can therefore solve the triangle. In view of the great uncertainty which exists as to the height of the atmosphere, we shall ascertain what the length of the third side of our triangle would be—the third side being the depth of the atmosphere in the line of sight towards the Sun on the horizon, according as the height of the atmosphere is 500 miles, 200 miles, 160 miles (being the mean of the figures given by Dr Buchan), or 50 miles.

We find by this simple method that, if the height of the atmosphere is 500 miles, the extent of the atmosphere in the line of sight towards the horizon is about 2,052 miles. If the height of the atmosphere is 200 miles, the extent of the atmosphere in the line of sight towards the horizon is nearly 1,275 miles, and if the height of the former is 160 miles, or 50 miles, the extent of the latter is about 1,137 miles, or 631 miles respectively. The extent of the atmosphere in our line of sight towards the horizon is thus enormously greater than it is towards the zenith, the actual extent varying from over four times, if the maximum estimate of the height of the atmosphere (500 miles) is correct, to nearly thirteen times, if one of the lowest estimates of the height is accepted.

This conclusion is arrived at by taking the Sun as the object of view. In the case of the Sun, or any other celestial body, excepting only the Moon, the angle formed by the line of sight from two different parts of the Earth, however far apart, is almost, or entirely, inappreciable. The size of the Earth is quite insignificant in comparison with the distance of the heavenly body. In the case of the Moon, however, this is not so. If we take the Moon, instead of the Sun, as the luminary observed from the respective positions, the line of sight towards the horizon will not be parallel to the line towards the zenith. It will slope towards the latter, so that the distance separating the two places of observation must be some-what less than a quadrant of the terrestrial circumference, and the angle formed at the centre of the Earth by two radii connecting the positions will be slightly less than a right angle. Notwithstanding this, however, the result as regards the extent of atmosphere through which the luminary is observed when seen on the horizon, in comparison with the extent of atmosphere through which it is observed when seen in the zenith, is the same in the case of the Moon as in the case of the Sun. Thus the extent of the atmosphere in our line of sight towards the Moon on the horizon, as towards the Sun in the same position, is about 2,052 miles, 1,275 miles, 1,137 miles or 631 miles, according as the actual height of the atmosphere is 500, 200, 160, or 50 miles.

Not only are we looking through a vastly greater

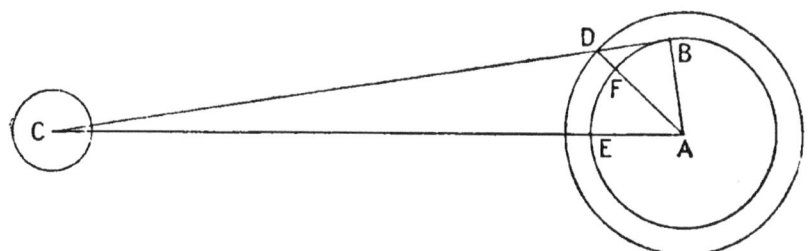

Diagram showing method of finding the comparative depth of the atmosphere in the line of sight towards the Moon on the horizon.

Small circle = The Moon.

Inner large circle = The terrestial equator.

Outer large circle = The outer boundary of the Earth's atmosphere.

A = The centre of the Earth.

C = The centre of the Moon.

B = Position at equator having the Moon on the horizon.

E =   „   „   „   „   in the zenith.

D = The upper boundary of the atmosphere in the line of sight from B towards the Moon on the horizon.

AB, AE and AF, being equatorial radii of the Earth, are each 3,963 miles in length.

The angle at B is a right angle.

To find the extent of the atmosphere in the line of sight from B towards the Moon on the horizon,—being the length of the line BD.

1. *If the height of the atmosphere is 500 miles,—*

AD will then be 3,963 miles + 500 miles, that is 4,463 miles.

$BD^2 = AD^2 - AB^2 = 4463^2 - 3963^2 = 19918369 - 15705369$
$= 4213000.$

BD = 2,052 miles.

Similarly, if the height of the atmosphere is taken to be 200 miles, 160 miles, or 50 miles, the extent of the atmosphere in the line of sight from B towards the Moon on the horizon will be found to be 1,275 miles, 1,137 miles or 631 miles respectively.

extent of atmosphere when we look towards a heavenly body at the horizon than when we look towards it in the zenith, or at the higher levels, but, in the horizontal view, an abnormal proportion of the atmosphere in our line of sight consists of the lower layers, and therefore of the heavier, and to some extent, probably, more humid particles. From this it comes about that the density of the atmosphere in the horizontal view is even greater than its depth in the line of sight might lead us to suppose.

These facts explain how the horizon is so frequently obscured by mists and fogs; how it is so exceptional, in the humid climate of Great Britain, to see the Sun or the Moon rise or set in a clear and cloudless sky.

We find, therefore, that there are two causes which conduce to the horizontal vision being atmospherically distinguished from the line of sight towards the higher altitudes. First:—The extent of the atmosphere in the horizontal line of sight is vastly greater than it is in the line of sight towards the greater altitudes. Second:—An abnormal proportion of the atmosphere in the horizontal line of sight consists of the lower and heavier layers.

Although, in measuring the comparative extent of the atmosphere in the horizontal line of view, we have made use of positions on the terrestrial equator, there is no reason to suppose that the extent of atmosphere towards the horizon differs according to either latitude or longitude. It seems most probable that, if the surface level is the same, the extent of atmosphere towards the horizon will be similar in all places on the Earth. Every place may be considered as being situated, to those who for the time occupy it, at what we may call "the top of the Earth." Every place, having an unobstructed view, is situated at the centre of its local atmosphere, in so far as the atmosphere rests on the Earth; that is to say, if the view all around is un-obscured, the lines of sight towards the horizon in the direction of every point of the compass, may be considered as radii of a circle.   Of course the length of radii may, and do, vary in different localities, and even in the same place at different times according to the clearness or obscurity of the atmosphere, but, so far as is known, they do not differ merely in respect of the position of the place on the surface of the Earth, so long as the height of the place, in relation to sea level is the same.   On the other hand, the extent of the atmosphere in the line of sight differs according to the height above sea level.   On the top of a high mountain, the extent of the atmosphere through which the line of sight towards the zenith passes, is considerably less than it is at sea level, and the extent of the atmosphere through which the line of sight towards the horizon passes is also different.   As the height of the atmosphere towards the zenith is reduced, the extent of the atmosphere in the line of sight towards the horizon is increased.   The latter is directed earthwards, and, just as an elevated position gives an enlarged view of the sur-rounding scenery, so it gives an increased extent of atmosphere in the earthward line of sight.   This is evident by a glance at our diagram illustrating the in-creased extent of the atmosphere in the line of sight with approach to the horizon.   It would thus appear that, if the increased extent of atmosphere in the line of sight towards the horizon is an inducing cause of the phenomenon with which we are dealing, the result, in similar conditions of the atmosphere, should be *more marked in observations from elevated positions than in observations from low levels.* There does not seem to be any evidence available bearing upon this aspect of the question.

Of course, in the latitudes where the Sun and Moon never occupy the position of being observable in the zenith, the contrast, as regards the extent of the atmosphere in the line of sight, between the higher altitudes attained by the Sun or Moon and the horizon, will not be so excessive as in the places where the luminary rises to the zenith. It will be seen, however, by reference to the last mentioned diagram, that the variation in the extent of the atmosphere in the line of sight is at first very slow as the zenith is departed from, and increases very rapidly as the horizon is approached.

We may now conclude that an apparent enlargement of heavenly bodies near the horizon does undoubtedly occur, and that its existence, or, at any rate, its amount, is dependent on causes of a variable character. We have seen that the extent of the atmosphere in the line of sight undoubtedly increases with approach towards the horizon, and that it is greatest just in the position occupied by our luminaries when they appear most enlarged. We know that the state of the atmosphere as regards humidity is of a constantly changing character, and we know that fog or humidity has an enlarging effect when objects of a luminous nature, close at hand, are perceptible through it. The changes which we have found to occur in connection with the atmosphere, both in its character and in relation to the point of view, appear, by analogy, to be exactly those required to furnish a reasonable explanation of the apparent enlargement of our luminaries when low in the heavens. In these circumstances, it would seem most probable that the changes in relation to the atmosphere are really the cause of the apparent changes in the heavenly bodies observed through it. Although the proof that this is so cannot be said to be absolutely complete, it is nevertheless of considerable weight. There is, therefore, fair reason for believing that the increased thickness of the atmospheric veil in the line of sight towards the horizon, in comparison with its thickness in the line of sight towards the higher altitudes, and the state of the constituent particles of the atmosphere as regards humidity, are, if not the only cause, at any rate the chief cause of the apparent enlargement which occurs to our luminaries; and of the corresponding changes in relation to the constellations as they sink towards the western, and as they rise above the eastern, horizon—a phenomenon, which for centuries, has been a subject of interest and speculation in the scientific world.

DOES THE WEATHER MOVE IN CYCLES?

SYNOPSIS.

Interest and importance of the weather—Antiquity of weather prognostications—Progress and present position of meteorology—Local character of weather conditions—Natural phenomena of a varying character usually subject to cycles—The Natural causes which influence the weather—Weather entirely dependent on state of atmosphere—Extent of the atmosphere—Conditions affecting the atmosphere—Comparison of the atmosphere to the waters of the ocean,—The tides and currents of the ocean and of the atmosphere—The Sun, the Moon, and the Earth's rotation the causes of the tides and currents in both sea and atmosphere—The various forms of lunar revolution—No exact recurrence of relations of Sun, Moon and Earth ever takes place—Approximate recurrence of their relations considered—Sidereal and synodical revolutions of the Moon—Nineteen years cycle of these revolutions in relation to the Earth's rotation—Multiples of the 19 years cycle considered—The cycles applicable to 19 years, 38 years, and 57 years compared—Fifty-seven years cycle considered in relation to the other forms of lunar revolution and to the other movements of the Earth—The Sun's rotation — Artificial causes affecting the weather—Summary and conclusion.

DOES THE WEATHER MOVE IN CYCLES?

Of the various phenomena of Nature there is probably no other of such general interest as the weather. To the agriculturist, to the mariner, to the holiday-maker, it is a subject of extreme importance; and these three are the most representative of classes, typifying industry, commerce, and pleasure. Even in this, the twentieth century, there is nothing of greater consequence to the welfare of humanity in all its diverse relations, than this threadbare subject of casual conversation—the weather.

To us, in Great Britain, the weather is a topic of never-failing interest. Our first duty as we meet acquaintances, after, perhaps, making indifferent enquiry as to their state of health, is to endeavour to describe the passing state of the weather. With us the weather is always new. In many parts of the Earth the weather is, to a great extent, of the same character during a prolonged period. To-day is as yesterday, and, in all probability, to-morrow will be as to-day. In these Islands we have a delightful uncertainty in this matter. It may, according to the calendar, be midwinter, but the weather of the day may possibly be that of midsummer, but if December in this way borrows a day from June we may take it as certain that it will not overlook making repayment six months' after date.

The weather being thus of supreme interest, it is natural that much attention should have been given to it by scientists. From the earliest times the discovery of the laws applicable to the weather has been a subject of profound study. Aristotle, who lived more than 300 years before our era, collected the popular prognostications of the weather which had accumulated before his time. Christ Himself reproves the Pharisees and the Sadducees for giving undue attention to this very subject,—" When it is evening ye say, 'it will be fair weather for the sky is red,' and in the morning, 'it will be foul weather to-day for the sky is red and lowering : O ye hypocrites, ye can discern the face of the sky, but can ye not discern the signs of the times ?"

The subject of meteorology has, in recent years, been most carefully studied, and great stores of records have been accumulated. The changes in temperature, atmospheric pressure, humidity of the air, and the direction and force of the wind throughout the varying seasons, have, in many parts of the Earth, received the most painstaking observance. Great advances have been made in our knowledge of the prevailing winds in the different seasons, and of the summer and winter mean temperature, over almost all the surface of the globe. Much information also has been gained as to the progressive motion of storms or cyclonic disturbances, and as to the spread of anti-cyclones with their generally light winds and clear atmosphere. The direction in which disturbances are travelling, their rate of progression, and whether they are increasing or diminishing in force are now very frequently predicted with wonderful correctness. Weather forecasts, for certain wide districts and for a limited time, have, indeed, attained much accuracy, and they may be accepted as being—at least to a considerable extent—true, on the mean for the district, in the vast majority of cases.

Although this is so, yet the general feeling in regard to the progress of meteorology, as its progess is shown by the weather forecasts issued by the Meteorological Office, is one of disappointment. We know that, although the forecasts are usually characterized by a considerable measure of accuracy, no reliance can be placed upon them. They may, or they may not, be correct. The fact that we are informed by the Meteorological Office that fine weather is to prevail on a certain day does not warrant our leaving our umbrellas at home. There is no certainty that the day will not be wet and disagreeable. Notwithstanding the fact that meteorology has for generations received so much study, and that the changing conditions of the weather have all been most carefully considered, we have, on the whole, excepting for the information obtained by the telegraph as to the prevailing weather in certain widely separated districts and the relative movements of the barometer and the thermometer, made but little advance since the time of the Pharisees and the Sadducees, who said in the evening "it will be fair weather for the sky is red;" and, in the morning, "it will be foul weather to-day, for the sky is red and lowering." The laws affecting the weather are still but little understood; meteorology is still very far from being an exact science.

One of the greatest difficulties—probably indeed the greatest difficulty—in connection with the forecasting of

the weather for any special place lies in the fact that
the weather bears to a large extent purely local
characteristics. The general average of the weather for
an extensive district may be foretold with some degree
of certainty, but locally the weather may differ materially
from the general average. The state of the weather is to
such an extent of a local character, that a perfect system
of forecasting would require to supply variations for
very limited areas. The weather may be comparatively
fine at one place, while, at exactly the same time, at a
place twenty miles away, it is wet and gusty. Yet a
forecast for a district including both places might
reasonably enough claim perfect accuracy if it described
the weather as "cloudy, variable, with some rain."

It will be seen that the present position of meteorological investigation is almost entirely observational, in
the hope that continued observation may by slow degrees
lead to a better comprehension of the laws applicable to
the atmospheric phenomena. At present we can only
vaguely anticipate the weather of to-morrow from the
signs of to-day, or, at best, the weather for the next few
days from the conditions prevailing over a wide district.
We cannot tell what conditions may afterwards come
into play; we cannot frame any general laws applicable
to the weather for any lengthened period.

Although this is the case, suggestions have from time to
time been made as to the existence of a weather cycle,—
suggestions usually based upon some supposed general
similarity in the character of the weather prevailing
throughout the district under consideration, in certain
different years, separated, perhaps, by a long interval.
These suggestions do not appear to be founded on any

scientific basis. The very fact, however, that the existence of a weather cycle has, from time to time, been suggested, goes to show the natural tendency of our minds to accept the existence of such a cycle.

It would certainly seem as if, in most natural phenomena of a varying character, a known cycle occurs. We have the cycle of the Earth's rotation in about twenty-four hours. We have the cycle of the Earth's orbital revolution in about three hundred and sixty-five days. We have similar cycles in the case of the planets. We have the cycles of the comets in their elliptical orbits. Even the whole solar system seems to have a cyclical motion around an unknown centre. We have the cycles of summer and winter, seed-time and harvest, day and night. We have the cycle of the tides in their daily rise and fall, the cycle of the Moon in its monthly journey around the Earth, the cycle of Sun spots in their recurring maxima and minima. We might further take illustrations from life itself with its cycles. We have the egg, the bird, the egg again; the seed, the plant, the seed again. It would seem as if almost all Nature's works have a cyclical character. It is, therefore, not unreasonable to conjecture that the weather also may conform to the same plan, and that there may therefore exist a cyclical period applicable to the weather with all its variations.

Let us now consider whether scientific observation affords any support to the supposition that the weather moves in cycles, or, at least, whether the existence of a weather cycle would be consistent with the results of scientific observation.

In this enquiry we have first to ascertain whether

there are any natural causes which influence the weather. On what does the weather depend?

It is reasonable to assume that the state of the weather is not absolutely accidental—that there must be certain causes in operation which result in the state of weather experienced in each separate locality from day to day, throughout all the varying changes of the year, whether the weather is hot or cold, wet or dry, stormy or calm, or intermediate between any of these states. Starting, then, with the assumption that the weather is dependent upon certain natural causes, let us endeavour to identify these causes and see how they operate.

It is, of course, manifest that the state of the weather, whatever it may be, depends entirely on the atmosphere. If the weather is warm, the atmosphere is heated. If the weather is dry, damp, or wet, the atmosphere is in varying degrees of humidity. If the weather is windy, the atmosphere is in more or less violent motion. If the weather is calm, the motion of the atmosphere is not so perceptible. Now, the atmosphere is simply the air which encircles us. We, on the surface of the Earth, are passing our lives at the bottom of a great ocean of air, just as certain fish may pass their lives at the bottom of the great ocean of water. The weather we experience is decided by the atmospheric conditions at the bottom of this great atmospheric ocean. Now the atmospheric conditions on the lowest level may be very different from the conditions at other levels. We may possibly be in a calm at the bottom of the atmospheric ocean, while, at a higher level, a storm may be raging. On the other hand, we may be in a storm, while the higher levels have a profound calm. As regards heat and cold, however, this interchangeability of condition does not seem to apply, or, at least, does not apply to the full extent. It would appear, from the evidence obtained by the aid of balloons, that, as a general rule, the temperature of the atmosphere falls as the elevation increases, just as it has been found, by mining and boring, that the temperature increases with the depth. Of course, the decrease of temperature with elevation above the land surface may not be regular. There are, no doubt, layers of air of different temperatures at varying levels, just as, in the sea, currents may be found at varying depths different in temperature from the adjoining water. We may, indeed, say that there are rivers in the sea; and, doubtless, there are also atmospheric streams of varying temperature in the great atmospheric ocean which surrounds the Earth. Again, the rule that the temperature decreases with elevation is subject to some slight modification in the case of elevated land surfaces. Although the tops of mountains are generally colder than the neighbouring valleys, yet, in the blaze of sunshine, there may be greater heat on the mountain top than in the valley. This is, however, temporary, and is compensated for by the more rapid radiation which occurs in the lighter air of the mountain top after the sunshine is withdrawn. It may thus be accepted that, as a general rule, the temperature of the atmosphere falls as the elevation from the sea level is increased.

There is some difference of opinion as to the depth or height of the atmospheric ocean at the bottom of which we pass our days. Dr. Alexander Buchan mentions that " from observations of luminous meteors, it has been inferred that the height (of the atmosphere) is at least 120 miles, and that, in an extremely attenuated form, it may even considerably exceed 200 miles." Other authorities have estimated the depth or height of the atmosphere variously, at from about 40 miles to 500 miles. It is agreed that, at a comparatively moderate distance from the surface of the Earth, the atmospheric pressure becomes very small. With each ascent of $3\frac{1}{2}$ miles, it is said that the density of the air is halved. As the atmospheric pressure at sea level is about $14\frac{1}{2}$ lbs. to the square inch, it will, according to this statement, at $3\frac{1}{2}$ miles above sea level, be about $7\frac{1}{4}$ lbs. to the square inch, at 7 miles, $3\frac{5}{8}$ lbs., and so on. The greatest height ever attained in the interests of science, appears to have been by a balloon sent up (apparently at Berlin) for the purpose of ascertaining the atmospheric pressure, and getting other information, for the purposes of meteorology, as to the atmospheric conditions at a great height. The balloon contained scientific instruments adjusted so as to secure the desired information, but had no living occupant. It was found, on its descent, that it had attained a height of $13\frac{1}{2}$ miles, and, that at that elevation, the height of the barometer was $1\frac{1}{2}$ inches. As the average height of the barometer at sea level is about 30 inches, and as this corresponds to an atmospheric pressure of about $14\frac{1}{2}$ lbs. to the square inch, the pressure at the height of $13\frac{1}{2}$ miles was only about one-twentieth of that at sea level, and was therefore only about $11\frac{3}{7}$ oz. to the square inch, —being actually less than would be the case were the statement that the pressure is halved for every $3\frac{1}{2}$ miles of elevation absolutely correct. It will be seen, from the rapid diminution of the atmospheric pressure as the height is increased, how extremely attenuated the air must be at the higher levels. It may, therefore, be conjectured that the atmosphere in the higher levels will be peculiarly responsive to the outer influences which may bear upon it. The hold of terrestrial gravitation is comparatively weakened, and the subjection to the outer influences proportionately strengthened.

Now, as the atmosphere is comparable to a great ocean surrounding the Earth, and, in many respects, has a noticeable analogy to the great ocean of water which rests on the surface of the Earth, we may, not unreasonably, observe how the waters of the ocean are influenced, and what are the causes which affect them, and judge whether the same causes may similarly influence the great atmospheric ocean.

The most noticeable occurrence in connection with the waters of the sea is the daily tidal motion. The causes of this motion have received much attention and they are well understood. The waters are heaped up in a long line of gradually lessening elevation from the centre to the extremities, extending from the Arctic regions to a diametrically opposite part of the Earth in the Antarctic regions, this long line being situated approximately on the longitudinal stretch of the Earth turned towards the Moon. The line extends not only along the whole length of the meridian nearly facing the Moon (which, of course, may not coincide in its entirety with any geographical meridian), through the direct attractive influence of the Moon on the waters, but also along the meridian nearly opposite, through the Moon attracting the solid mass of the Earth to a greater extent than it attracts the waters turned away from the Moon. From this it is evident that the tidal elevation really forms a kind of irregular ring around the Earth. The ring is, of course, broken here and there, by intervening lands rising above the surface of the water, and also (from the fact that the heaping up of the waters occupies a certain definite time), by the tidal elevation on the opposite sides of the Earth being somewhat deflected in opposite directions.

The general effect, however, is clear, and that is, that, through the influence of the Moon, a heaping up of the water occurs in the form of a circle, or rather two semi-circles, encompassing the Earth approximately on the meridian for the time being under the Moon, and the opposite meridian. Through this irregular or broken ring, the Earth turns in each daily rotation; and the ring itself revolves around the Earth once a month, following the Moon.

The Sun has a similar influence on the waters, but its influence is much weaker than that of the Moon. When the Sun and the Moon act in harmony, as at new Moon and full Moon, the attractive power of the Sun is added to that of the Moon and the elevation of the waters is consequently increased. When they act in opposition, as at first quarter and third quarter (that is half-moon), the elevation of the waters is lessened by the drawing away of water by the Sun. In the former case we have the spring or high tides with very low tides intervening. In the latter case we have the neap tides, with their comparatively moderate rise and fall. In both cases the high water is associated with the Moon. When the Sun's influence is in conjunction with that of the Moon, then the Sun is associated with the high tide; when the Sun is in opposition to the Moon then the Sun is associated with the low tide,—the separate influence of the Sun

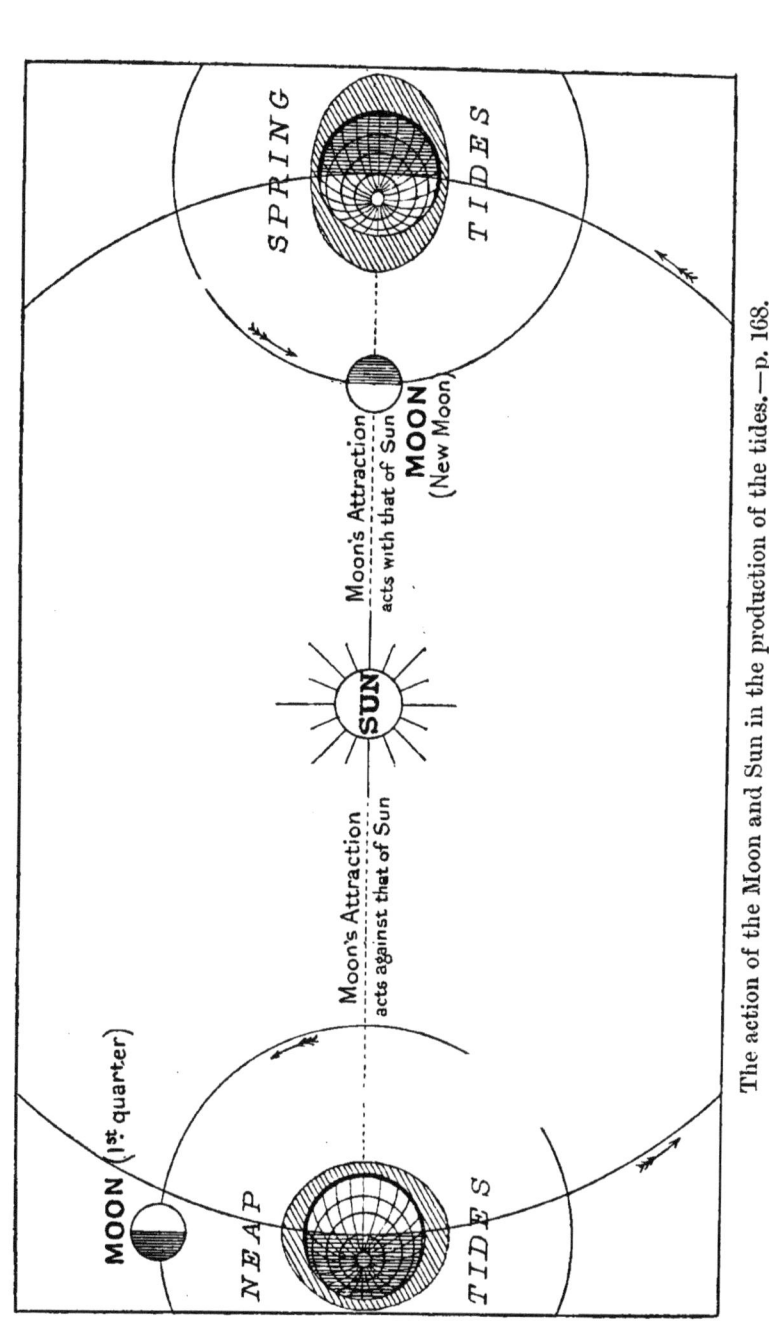

The action of the Moon and Sun in the production of the tides.—p. 168.

having merely the effect of preventing the ebb-tide being so low as it would otherwise be. The Moon primarily, and the Sun subordinately, are the causes of the tides, and the production of the tides is the most noticeable of the effects produced on the waters of the ocean.

While the Moon has thus the chief influence on the tides, the Sun has, irrespective of the Moon, a great influence on the waters. Thus, the unequal heating of the water in different portions of the sea causes currents and varying degrees of evaporation. In short, the waters of the ocean are influenced by the Sun and the Moon in various ways; and the Sun and the Moon together are, in combination with the rotation of the Earth, the cause, either directly or indirectly, of almost all the changes which take place in connection with the seas.

If then, the Sun and the Moon, in association with the Earth's rotatory motion, exercise such a stupendous influence on the waters of the ocean as to cause them to rise up in a heap to the height of perhaps as much, in certain places, as 40 feet or 50 feet (that being the occasional difference in depth between high water and low water), and to rush in currents, hither and thither, for, perhaps, thousands of miles through the main body of the waters, we may readily conclude that they must have an equal, or greater, influence on the vast ocean of the air which forms the terrestrial atmosphere.

The weight of the air at sea level is equivalent on the average to the weight of a column of water $32\frac{1}{2}$ feet in height; that is to say, a cylinder of air extending from sea level to the limits of the Earth's atmosphere would weigh the same as a cylinder of water, of the same circumference, 32½ feet long, the actual weight being, say, 14·6 lbs. to the square inch, The cylinder of air would, however—if we accept the mean of the figures given by Dr. Alexander Buchan as fairly representing the extent of the atmosphere—be about 160 miles in length.

In regard to the unequal heating arising from the difference of exposure to the solar rays, and the consequent currents, a distinction appears to exist between the sea and the atmosphere. The greatest reception of heat by the waters of the ocean occurs at the highest level, the superficial layer being that which is chiefly heated. With the atmosphere, as we have noticed, the opposite appears to be the case. The evidence afforded by balloons goes to show that, generally speaking, the temperature falls as the height increases, the highest temperature being, on the mean, at or near the surface. It is believed that the atmosphere is heated mainly by conduction from the surface of the Earth, the heat being then communicated from particle to particle of the air. Thus the greatest heat occurs at and near the surface, which is where the air is most dense and the atmospheric pressure greatest. The air, as it becomes heated, rises to a higher level, but, as it ascends, it gradually parts with its heat.

It is, however, unnecessary, in our present enquiry, to speculate as to the methods of atmospheric action by which weather changes are brought about. All that is required for our purpose is to ascertain what are the *causes* which produce the atmospheric actions which result in these changes; and what are the relative changes in these causes themselves. We have seen that, in all probability, the causes are the Sun and the Moon in association with the diurnal rotation of the Earth. Let us now consider the relation of the Sun and the Moon and the Earth's rotation to each other.

It is evident that, if the causes which influence the weather are, as we have supposed, the Sun and the Moon and the Earth's rotation, then the weather must repeat itself when these three return to an exactly similar relation. Is there then any likelihood of such a return, and, if so, what period will elapse in the interval? Such a period would be in reality the weather cycle.

The essentials necessary to secure the return of the conditions which we have supposed to have the effect of deciding the weather are,—(1) that the Earth is again, in its daily rotation, in exactly its former position in relation to the Sun; (2) that the Earth has returned to exactly the same part of its orbit; and (3) that the Moon has returned to exactly the same position in relation to both the Sun and the Earth. We might add a fourth condition, which would be that the Sun itself had returned to exactly the same stage in its axial rotation, so that the same part of the Sun would be bearing upon the Earth. This condition is, however, of uncertain importance, and may, possibly, be of no consequence. The period of the *Moon's* rotation corresponds with the period of its revolution around the Earth, so that the Moon will necessarily have the same part bearing on the Earth on its return to the same part of the heavens. We may therefore confine our attention to the three essentials we have mentioned.

As the mean period of the Earth's diurnal rotation in relation to the Sun (for instance from noon to noon), is exactly twenty-four hours, it is necessary, in order to secure the second and third essentials, that they should form an exact multiple of twenty-four hours.

We shall now consider the motions of the Moon.

The revolution of the Moon around the Earth is, of course, the origin of our month; but the month is, from this point of view, of uncertain length, as there are different kinds of lunar revolution, and the length of the month will depend on the nature of the lunar revolution which is taken as its basis.

We have, first, the sidereal revolution,—which is the time occupied by our satellite in passing from a star back to the same star again. As the star may be considered as fixed, and as the change of position of the Earth in relation to the star may be disregarded, this may be taken as the true period of the Moon's revolution. Its length is 27·32166 days.

We have, next, the anomalistic revolution of the Moon, by which is meant its return to the same part of its orbit. The distance of the Moon from the Earth is subject to considerable variation, varying from about 221,593 miles when in perigee, or nearest to the Earth, to about 252,948 miles when in apogee, or farthest away from the Earth. The orbit of the Moon is not constant, and, consequently, the Moon, on completing a revolution (that is, after travelling 360 degrees from where it started), has not returned to exactly the same part of its orbit. We may, for the moment, regard the orbit as if it were composed of a material substance. In the time the Moon is describing the revolution, the orbit may be said to have moved round a little way in the same direction as the Moon is travelling. Thus, if we take the revolution as commencing at perigee, or the part of the orbit nearest to the Earth, we find that, during the time the Moon is travelling around the heavens, perigee has itself moved on a small distance, and the Moon has to travel this small distance more than a revolution (that is, 360°) before it is again in perigee. The period of the anomalistic revolution is 27·55460 days, which is ·23294 of a day longer than the sidereal revolution.

We have, next, the tropical revolution, which is the time it takes the Moon to return to the same celestial longitude, or right ascension. Right ascension is measured from the part of the heavens at which the Sun crosses the equator at the vernal equinox. Let us suppose the Sun to be in the zenith at, say, the point where the meridian of Greenwich intersects the equator, in a certain year in the month of March. That position of the Sun marks the moment of the vernal equinox, and it also marks the hour of noon in Greenwich time. A year later, however, the Sun will be in the zenith on the equator not at the point intersected by the meridian of Greenwich, but at a point considerably to the west of the meridian of Greenwich. When the Sun crossed the meridian of Greenwich immediately before this crossing of the equator, it was in the zenith a little to the south of the equator, and when it next crosses that meridian it will be in the zenith a little to the north of the equator. As the crossing of the meridian of Greenwich by the Sun marks the hour of noon in Greenwich time, and, as the time in Greenwich is later than noon when the Sun is west of that meridian, it follows that, with the Sun crossing the equator considerably to the west of the meridian of Greenwich, the vernal equinox will occur considerably later than noon in Greenwich time. The place in the heavens which the Sun occupies, when it appears in the zenith at the terrestrial equator in its northward course, is the place from which right ascension, or celestial longitude, is measured; and, in each succeeding year, this place is slightly to the west of where it was in the preceding year. As the Sun's annual course is, to our view, eastward among the stars, it follows that, as the Sun crosses the equator in each succeeding year to the west of the point at which it crossed in the preceding year, the point of intersection slightly *precedes* the completion of the apparent solar revolution. This change of the point of intersection of the Sun's apparent path and the celestial equator, is called the precession of the equinoxes, and the cause is understood to lie in the Poles of the Earth's axis describing a circle—in a period of nearly 26,000 years—whereby, in each succeeding year, they are pointed to a slightly different part of the heavens, and the terrestrial equator is proportionately displaced in relation to the heavens. The amount of the annual precession is rather more than 50 seconds of the arc of the circle—say 50·28 seconds—and, as it takes the Sun about 365 days, 6 hours, 9 minutes, 9 seconds, to complete its apparent annual sidereal revolution eastward around the heavens, it would take it about 20 minutes, 24 seconds in time to travel over 50·28 seconds of the arc of the circle. The Sun is, therefore, in the zenith at the equator about 20 minutes, 24 seconds, before, in its apparent eastward journey, it completes a revolution of the heavens.

Now, in the sidereal year (365 days, 6 hours, 9 minutes, 9 seconds), the Earth makes rather more than 365¼ rotations on its axis, so that, on the completion of the Sun's apparent annual revolution, the meridian facing the Sun would not be the meridian which faced it at the commencement of the revolution,—which was, we have supposed, the meridian of Greenwich—but the meridian more than a quarter of the circumference of the Earth west thereof,—being, say, the meridian 92° 17′ 15″ W. As we have seen, however, the Sun crosses the equator slightly before completing an apparent revolution; its place in the heavens, at the time of cross-ing the equator always *preceding*,—that is, being slightly to the west of,—the point at which it crossed in the previous year. Therefore the Earth would not have rotated quite so much as this when the crossing next occurs. The Earth would require about 20 minutes, 24 seconds, more to do so, and, as four minutes of time are equivalent to a change of one degree in the meridian facing the Sun, 20 minutes, 24 seconds, would represent 5 degrees, 6 minutes. If, therefore, we subtract 5 degrees, 6 minutes, from 92° 17′ 15″ W., we get, approximately, the meridian above which the Sun crosses the equator in its apparent northward course in the year follow-ing the crossing at the meridian of Greenwich. The meridian will thus be about 87° 11¼′, W. of that of Greenwich. Thus the point at which the Sun is in the zenith at the equator is, at each succeeding spring equinox, about 87° 11¼′ to the west of the point at which this occurred in the preceding year.

The slight annual displacement (50·28 seconds), in the point from which right ascension, or celestial longitude, is measured, becomes still less when it is regarded from the point of view of the lunar revolution. A displace-

ment of 50·28 seconds a year will be only about 4 seconds
in a lunar revolution. Its effect is to make the period
of the tropical revolution of the Moon 27·32156 days,
which is ·0001 of a day less than the period of the
sidereal revolution.

There still remain two forms of the lunar revolution,
—the nodical and the synodical.

The nodical revolution is the passage from the point
at which the Moon's orbit, to our view, intersects the
Sun's apparent course back to the same point again.
When the Sun and the Moon are, at the same time, at, or
within a certain distance of, these points of intersection,
a total or partial eclipse occurs. The nodes of the Moon's
orbit are not fixed. If, as we did when considering the
anomalistic revolution, we regard the Moon's orbit as
consisting of a material substance, we may say that,
while the Moon is revolving around the Earth, the orbit is
moved back a little. Thus the Moon reaches the same
node for the second time a little before completing its
revolution. The time occupied by the Moon in passing
from a node back to the same node again is 27·21222
days, which is ·10944 of a day less than the period of the
sidereal revolution.

Most important of all the different kinds of lunar revolution is the synodical revolution, which is the revolution in relation to the Sun, in our line of view,—the time
between new Moon and new Moon, full Moon and full
Moon. It, of course, varies to a slight extent owing to
the varying speed of the Earth in its orbit according to
its distance from the Sun, in consequence of which we,
during the diurnal rotation, travel over a larger arc of
the orbital circle when nearer to the Sun than when

farther away. The average time of the synodical revolution is 29·53059 days. This is 2·20893 days longer than the time of the sidereal revolution. The difference is occasioned by the displacement of the Sun, caused by the orbital progression of the Earth during the period of the Moon's revolution. Thus the Moon has to make more than a complete revolution before it again gets into the same relation to the Sun. The time occupied in the sidereal revolution is as nearly as possible the true time of the Moon's revolution, but, as the Sun is the centre of our system and the source of our time-reckoning, the synodical revolution is the most important to the inhabitants of the Earth.

The different forms of the lunar revolution are therefore as follows :—

Sidereal Revolution	(star to star)	27·32166 days.
Anomalistic ,,	(perigee to perigee)	27·55460 ,,
Tropical ,,	(equinox to equinox)	27·32156 ,,
Nodical ,,	(node to node)	27·21222 ,,
Synodical ,,	(new moon to new moon)	29·53059 ,,

As there is a trifling discrepancy between different authorities in regard to some of these periods it is right to mention that the times given for the various forms of lunar revolution as above—on which our calculations are based—are those specified by the late Mr R. A. Proctor, the distinguished authority on astronomical matters. Of course it must be kept in mind that, in large calculations, discrepancies may arise from the fact that the periods are given to no more than five decimal places.

As our purpose is to endeavour to ascertain whether the Sun and the Moon return to exactly the same

relation to each other, and to the same part of the Earth
and the heavens after any certain period, and, if so,
what the period is, it would seem to be sufficient for our
purpose to confine our attention more particularly to the
sidereal and synodical revolutions. The former marks
the return of the Moon to the same part of the sky, and
the latter marks its return to the same position in
relation to the Sun. Can we calculate any period for
the recurrence of these relations in association with the
same portion of the earth? Supposing the Sun, the
Moon, and a certain star, to be on the meridian of
Greenwich at a certain moment, at certain elevations,
when, if ever, shall all these relations again coincide?

It would appear that nearly one hundred thousand
millions of revolutions (98,435,300,000 sidereal revolutions or 91,072,200,000 synodical revolutions), will have
to occur before the relations are exactly repeated. As
this would occupy a number of years extending into ten
figures, during which it cannot be doubted that changes
would have occurred in relation to the Earth, and even
in relation to the heavens, upsetting the correspondence,
it may be concluded that no exact return of the
conditions ever occurs.

Although this is so, there would appear to be some
reason to suppose that an approximation to the relations
does occur, and that within a period of human interest.
An approximate return of the relations of the various
causes affecting the weather may be supposed to occasion
an approximate return of the weather conditions.

Let us, then, examine more particularly the periodical
relations of the sidereal and synodical lunar revolutions
in relation to the diurnal rotation of the Earth.

DIAGRAM SHOWING THE VARYING DIVERGENCE IN THE TIME OF COMPLETION OF SERIES OF SIDEREAL AND SYNODICAL REVOLUTIONS OF THE MOON WHICH COMMENCE AT THE SAME TIME

(Up to 762 Sidereal Revolutions, being 57 Series.)

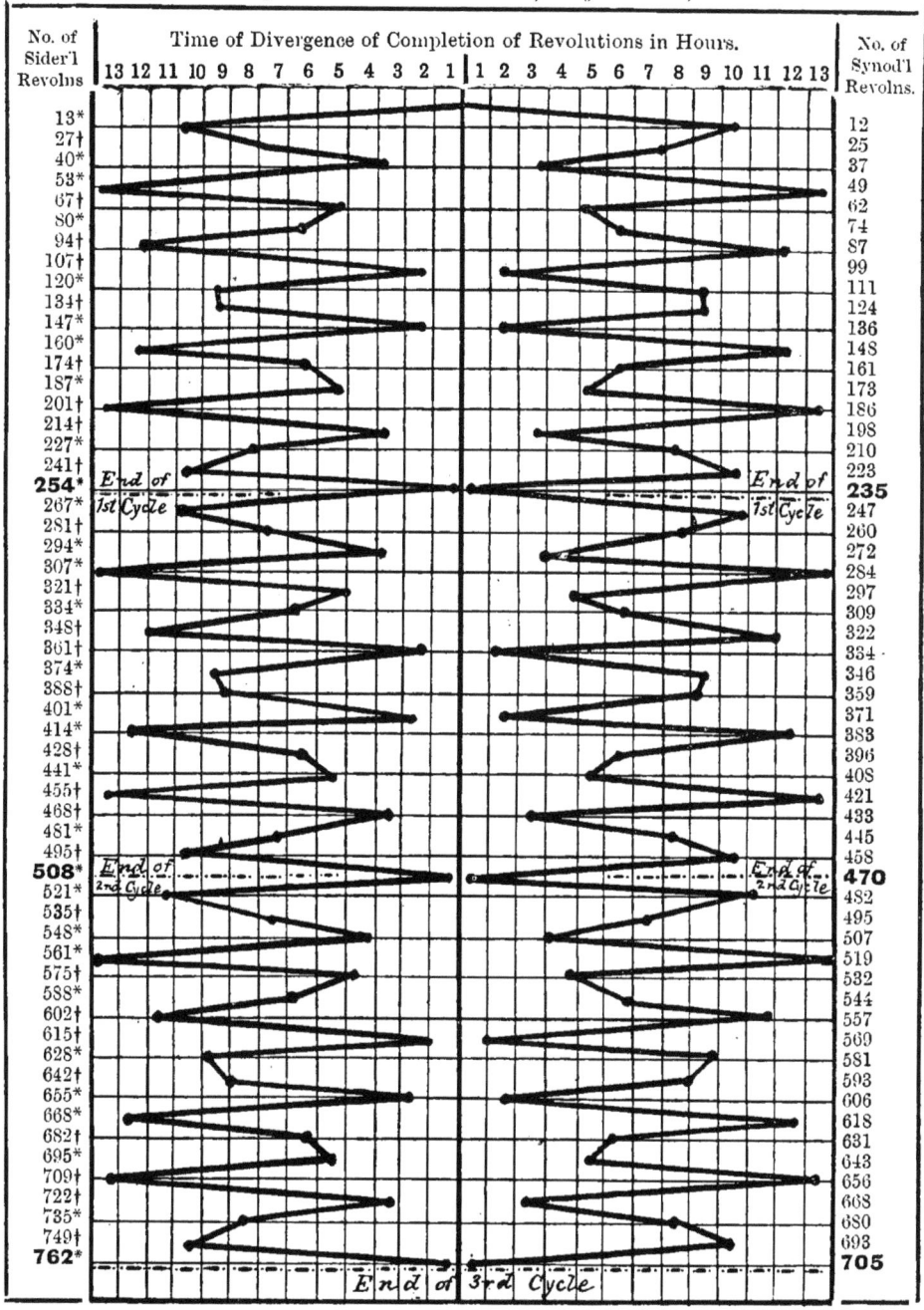

* † At the numbers marked with an asterisk (*) the Sidereal Revolutions are completed *before* the corresponding Synodical Revolutions; while at the numbers marked with a dagger (†) the Sidereal Revolutions are not completed till *after* the corresponding Synodical Revolutions. This results, in some instances, in ensuing series which appear by the diagram to be very similar in the nature of their divergence, being, in reality, very dissimilar. It will be noticed that the rotation of asterisks and daggers is exactly the same in each cycle.

The difference between the time of a revolution of the Moon in relation to a star (27·32166 days), and a revolution in relation to the Sun (29·53059 days) is, as we have seen, 2·20893 days. On this basis, the following statement shows the relations between these revolutions for a limited period.

TABLE SHOWING THE VARYING DIVERGENCE IN THE TIME OF COMPLETION OF SIDEREAL AND SYNODICAL REVOLUTIONS OF THE MOON WHICH COMMENCE AT THE SAME TIME (UP TO 14 SIDEREAL REVOLUTIONS).

1	Sidereal Revolutn.	+	2·20893	days	=	1	Synodical Revoln.
2	„ Revolns.	+	4·41786	„	=	2	„ Revolns.
3	„ „	+	6·62679	„	=	3	„ „
4	„ „	+	8·83572	„	=	4	„
5	„ „	+	11·04465	„	=	5	„
6	„ „	+	13·25358	„	=	6	„
7	„ „	{ + 15·46251	„	= { 7	„
		{ − 14·06808	„	= { 6	„
8	„ „	−	11·85915	„	=	7	„
9	„ „	−	9·65022	„	=	8	„
10	„ „	−	7·44129	„	=	9	„
11	„ „	−	5·23236	„	=	10	„
12	„ „	−	3·02343	„	=	11	„
13	„ „	−	0·81450	„	=	12	„
14	„ „	+	1·39443	„	=	13	„ „

It will at once be noticed that the maximum difference between the time of completion of revolutions of the two classes occurs in the comparison between seven sidereal revolutions, and either seven or six synodical revolutions. This is the turning point. The closest correspondence between the two classes of revolution, on the other hand, occurs in the relation between thirteen sidereal revolutions, and twelve synodical revolutions. In the former case the difference

is from about fourteen days to over fifteen days. In
the latter case the separating period is only ·81450 of a
day. It will be seen that these periods mark,
respectively, the opposite seasonal extremes, and the
return of the same season. If we suppose the two
classes of revolution to commence on 1st January, then,
in the case where the greatest separation occurs, we
have passed to some time about the 1st of July; in the
case where the least separation occurs, we have returned
to some time about 1st January again.

Even, however, when there is the least separation, it is
still considerable. It is ·81450 of a day, which is equal
to over $19\frac{1}{2}$ hours. We have now to endeavour to
ascertain whether a smaller amount of separation
between the two kinds of revolution is of a recurring
character. As the annual return to a comparatively
small amount of separation marks the return of the
same *seasonal* character of weather, so a periodical
return to a less amount of separation should indicate the
periodical return of weather with a closer resemblance
than characterizes the recurring seasons.

We have seen that it is on the completion of thirteen
sidereal revolutions that the separation is least. We
shall now therefore take multiples of 13 sidereal revolutions, and, where the separation amounts to more than
one-half of the difference (2·20893 days), between one
sidereal revolution and one synodical revolution, we
shall add, or subtract, as the case may require, one
revolution, so as invariably to secure the least separation.
The difference must therefore in no case exceed 1·10447
days.

We have found that, on the completion of *thirteen*

sidereal revolutions, the Moon will have passed the point of completion of twelve synodical revolutions by ·81450 of a day; and, on completing *fourteen* sidereal revolutions, it will be 1·39443 days short of completing 13 synodical revolutions. We can therefore use these two periods (·81450 of a day and 1·39443 days), in ascertaining the times of closest approximation of complete revolutions.

Supposing the Moon commences a sidereal and a synodical revolution at the same moment, the result as regards complete revolutions of each class will be seen from the table on page 122.

We have here very clear evidence of a cycle in these revolutions. The time by which the completion of the Moon's 254th sidereal revolution is separated from the time of the completion of its 235th synodical revolution is only ·01299 of a day—that is about 18 minutes, 42·33 seconds. This interval is doubled on the completion of the 508th sidereal revolution and 470th synodical revolution, and trebled on the completion of the 762nd sidereal revolution and 705th synodical revolution, and so on. Thus, the time-interval on the completion of the first series (supposing we start with no time-interval), is about 18 minutes, 42·33 seconds; and on the completion of the second series, *twice* 18 minutes, 42·33 seconds, as compared with the commencement, but only 18 minutes, 42·33 seconds, as compared with the end of the first series; and, on the completion of the third series, *three times* 18 minutes, 42·33 seconds, as compared with the commencement, but only 18 minutes, 42·33 seconds, as compared with the end of the second series.

It will be noticed also that each series is exactly

TABLE SHOWING THE VARYING DIVERGENCE IN THE TIME OF COMPLETION OF SERIES OF SIDEREAL AND SYNODICAL REVOLUTIONS OF THE MOON, WHICH COMMENCE AT THE SAME TIME (UP TO 762 SIDEREAL REVOLUTIONS, BEING 57 SERIES).

	Sidereal Revolutions	±	days	=	Synodical Revolns.	
(0)	0 Sidereal Revolutions ±		·00000 days =		0 Synodical Revolns.	
(13) 13	,, ,,	−	·81450	,,	12	,, ,,
(14) 27	,, ,,	+	·57993	,,	25	,, ,,
(13) 40	,, ,,	−	·23457	,,	37	,, ,,
(13) 53	,, ,,	−	1·04907	,,	49	,, ,,
(14) 67	,, ,,	+	·34536	,,	62	,, ,,
(13) 80	,, ,,	−	·46914	,,	74	,, ,,
(14) 94	,, ,,	+	·92529	,,	87	,, ,,
(13) 107	,, ,,	+	·11079	,,	99	,, ,,
(13) 120	,, ,,	−	·70371	,,	111	,, ,,
(14) 134	,, ,,	+	·69072	,,	124	,, ,,
(13) 147	,, ,,	−	·12378	,,	136	,, ,,
(13) 160	,, ,,	−	·93828	,,	148	,, ,,
(14) 174	,, ,,	+	·45615	,,	161	,, ,,
(13) 187	,, ,,	−	·35835	,,	173	,, ,,
(14) 201	,, ,,	+	1·03608	,,	186	,, ,,
(13) 214	,, ,,	+	·22158	,,	198	,, ,,
(13) 227	,, ,,	−	·59292	,,	210	,, ,,
(14) 241	,, ,,	+	·80151	,,	223	,, ,,
(13) 254	,, ,,	−	·01299	,,	235	,, ,,
(13) 267	,, ,,	−	·82749	,,	247	,, ,,
(14) 281	,, ,,	+	·56694	,,	260	,, ,,
(13) 294	,, ,,	−	·24756	,,	272	,, ,,
(13) 307	,, ,,	−	1·06206	,,	284	,, ,,
(14) 321	,, ,,	+	·33237	,,	297	,, ,,
(13) 334	,, ,,	−	·48213	,,	309	,, ,,
(14) 348	,, ,,	+	·91230	,,	322	,, ,,
(13) 361	,, ,,	+	·09780	,,	334	,, ,,
(13) 374	,, ,,	−	·71670	,,	346	,, ,,
(14) 388	,, ,,	+	·67773	,,	359	,, ,,
(13) 401	,, ,,	−	·13677	,,	371	,, ,,
(13) 414	,, ,,	−	·95127	,,	383	,, ,,
(14) 428	,, ,,	+	·44316	,,	396	,, ,,
(13) 441	,, ,,	−	·37134	,,	408	,, ,,
(14) 455	,, ,,	+	1·02309	,,	421	,, ,,
(13) 468	,, ,,	+	·20859	,,	433	,, ,,
(13) 481	,, ,,	−	·60591	,,	445	,, ,,
(14) 495	,, ,,	+	·78852	,,	458	,, ,,
(13) 508	,, ,,	−	·02598	,,	470	,, ,,
(13) 521	,, ,,	−	·84048	,,	482	,, ,,
(14) 535	,, ,,	+	·55395	,,	495	,, ,,
(13) 548	,, ,,	−	·26055	,,	507	,, ,,
(13) 561	,, ,,	−	1·07505	,,	519	,, ,,
(14) 575	,, ,,	+	·31938	,,	532	,, ,,
(13) 588	,, ,,	−	·49512	,,	544	,, ,,
(14) 602	,, ,,	+	·89931	,,	557	,, ,,
(13) 615	,, ,,	+	·08481	,,	569	,, ,,
(13) 628	,, ,,	−	·72969	,,	581	,, ,,
(14) 642	,, ,,	+	·66474	,,	593	,, ,,
(13) 655	,, ,,	−	·14976	,,	606	,, ,,
(13) 668	,, ,,	−	·96426	,,	618	,, ,,
(14) 682	,, ,,	+	·43017	,,	631	,, ,,
(13) 695	,, ,,	−	·38433	,,	643	,, ,,
(14) 709	,, ,,	+	1·01010	,,	656	,, ,,
(13) 722	,, ,,	+	·19560	,,	668	,, ,,
(13) 735	,, ,,	−	·61890	,,	680	,, ,,
(14) 749	,, ,,	+	·77553	,,	693	,, ,,
(13) 762	,, ,,	−	·03897	,,	705	,, ,,

DIAGRAM SHOWING THE VARYING DIVERGENCE IN THE TIME
OF COMPLETION OF SIDEREAL AND SYNODICAL REVOLU-
TIONS OF THE MOON WHICH COMMENCE AT THE SAME
TIME. (*Up to* 14 *Sidereal Revolutions.*)

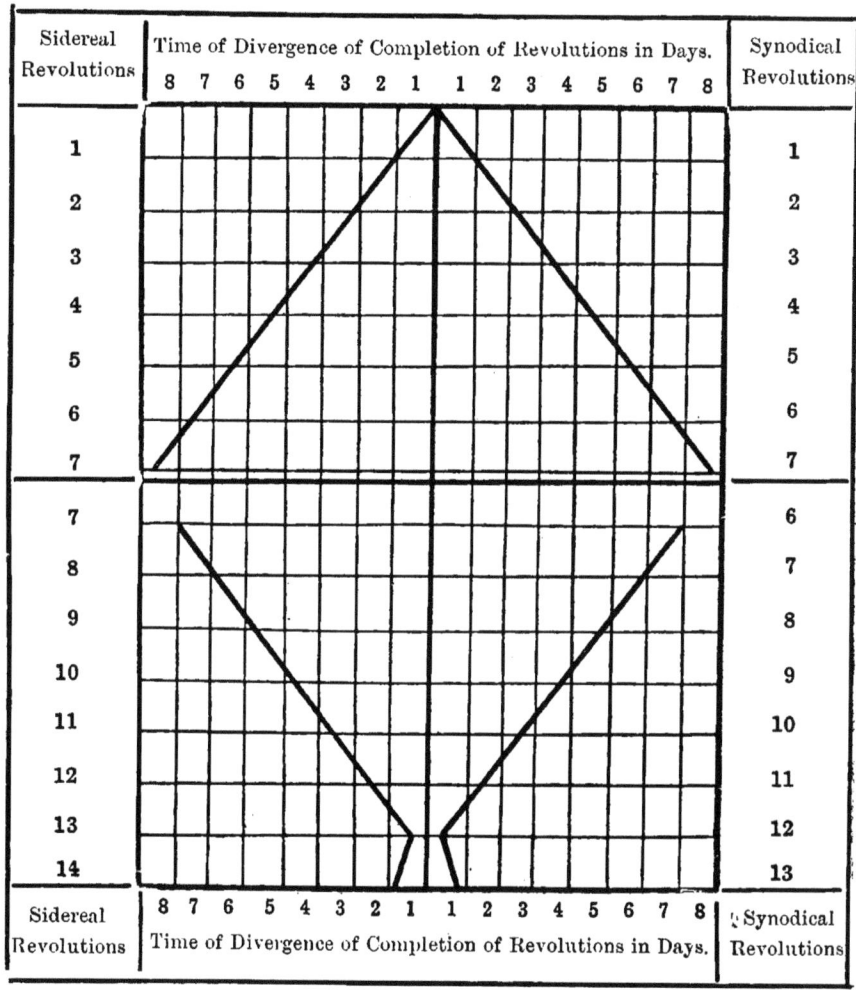

identical in its components. Thus the sidereal revolutions composing each series in succession are as follows, —each column being taken in its order.

13	13	13	13
14	14	13	13
13	13	14	14
13	13	13	13
14	14	14	254

We may now accept the conclusions shown in the following Table :—

TABLE SHOWING THE REGULAR PROGRESSIVE DIVERGENCE IN THE TIME OF COMPLETION OF CYCLES OF SIDEREAL AND SYNODICAL REVOLUTIONS OF THE MOON, WHICH COMMENCE AT THE SAME TIME (UP TO 2032 SIDEREAL REVOLUTIONS, BEING 8 CYCLES).

254	Sidereal Revolutions	− ·01299	of a day =	235	Synodical Revolution
508	,,	,,	− ·02598	,,	470 ,, ,,
762	..	..	− ·03897	,,	705 ,,
1016	..	..	− ·05196	,,	940 ,,
1270	..	..	− ·06495	,,	1175 ,,
1524	,,	,,	− ·07794	,,	1410 ,,
1778	..	,,	− ·09093	,,	1645 ,,
2032	,,	,,	− ·10392	,,	1880 ,, ,,

We have now to see how this fits in with the diurnal and annual revolutions of the Earth. The diurnal rotation decides the part of the Earth facing the Sun and Moon in their relative positions in the revolutions we have been considering, and, as we have seen, this is of the utmost importance—in view of the fact that the weather is essentially of a local character. The annual revolution is of equal importance, as deciding whether the cyclical periods coincide with the return of the Earth to the same part of its orbital course, and therefore to the same seasonal conditions.

As the sidereal revolution of the Moon occupies 27·32166 days, 254 sidereal revolutions will occupy 6939·70164 days; and as the synodical revolution occupies 29·53059 days, 235 synodical revolutions will occupy 6939·68865 days. This is, as nearly as possible, the number of days in 19 years, the number being 6939 days if the period contains only four leap-years, and 6940 days if it contains five leap-years. It will contain either four or five leap-years according to the date at which it commences. If we commence with the year 1901, the period of 19 years will commence with three common years and end with three common years, and will contain fifteen common years and four leap-years, amounting to 6939 days; but, if we commence with the year 1902, we shall have only two common years at the commencement before a leap-year, and we shall end with a leap-year, so that the period will contain 6940 days. Thus 254 sidereal revolutions of the Moon, and 235 synodical revolutions occupy, as nearly as possible, exactly 19 years. Therefore it is undoubtedly the case that the Earth is in practically the same part of its orbit at the end of the period as at the beginning.

Unfortunately, in regard to the diurnal rotation of the Earth, the position of matters is different. On the completion of 254 sidereal, and 235 synodical, revolutions of the Moon, the meridians facing the Moon are very dissimilar from the meridian which faced it at the commencement. We have, in this matter, to deal with the solar, and not the sidereal, diurnal rotation—that is, with a period, on the average, of 24 hours, not of 23 hours, 56 minutes, 4 seconds. In order that the stage of the diurnal rotation should be the same as at the commencement, the revolutions would require to occupy an exact number of days, instead of 6939 days with a considerable fraction over. On the completion of 235 synodical revolutions in 6939·68865 days, the meridian facing the Moon—supposing ? that at the start the meridian in that position was the meridian of Greenwich —will be, not the meridian of Greenwich, but the meridian lying 247° 54' 50·4" to the west of the meridian of Greenwich—that is 112° 5' 9·6" to the east of that meridian. Similarly, on the completion of 254 sidereal revolutions of the Moon in 6939·70164 days, the meridian facing the Moon will be 252° 35' 25·4" W., or 107° 24' 34·6" E., of the meridian of Greenwich. The distance of separation of the meridians facing the Moon on the completion of the respective classes of revolution is thus 4 degrees, 40 minutes, 35 seconds.

As, in each hour, the meridian facing the Sun is changed 15 degrees to the west through the Earth's diurnal rotation eastwards, it would require only 18 minutes, 42·33 seconds, to affect a change of 4° 40' 35". Therefore the Sun would be on the meridian 107° 24' 34·6" E., in that brief interval after being on the meridian 112° 5' 9·6" E. The interval is, in fact, ·01299 of a day, which we have found to be the time-separation between the completion of the 254th sidereal, and 235th synodical, revolution of the Moon.

Our purpose, however, is not to study the relations between the Sun, Moon, and star on the completion of the 19 years cycle far to the east of the meridian of Greenich, but to ascertain their relations when the Sun is once again on the Greenwich meridian.

We have seen that the synodical cycle is completed in 6939·68865 days, and that the Sun is then on the meridian 112° 5' 9·6" east of Greenwich. It will there-fore be ·31135 of a day before the Sun is again on the meridian of Greenwich, that being the difference between 6939·68865 days and 6940 days, and also the time it takes the Earth, by its daily rotation, to change the meridian facing the Sun to the extent of 112° 5′ 9·6″ westwards. The time represented by ·31135 of a day is 7 hours, 28 minutes, 20·64 seconds.

The time required for one synodical revolution is, as we have seen, 2·20893 days longer than the time required for one sidereal revolution. We may express this differently by saying that, during the time of the synodical revolution, which is 29·53059 days, the point in its orbit at which the Moon completes a sidereal revolution, apparently falls behind by the space° which the Moon travels over in 2·20893 days. This being so, we have to ascertain to what extent the same point of the lunar orbit will apparently fall behind in 7 hours, 28 minutes, 20·64 seconds, which is the time required to bring the meridian of Greenwich to face the Sun after the completion of the 19 years synodical cycle. This is merely a question of proportion. As the time of the synodical revolution is to 2·20893 days so is 7 hours, 28 minutes, 20·64 seconds to the required time. The time is 33 minutes, 32 seconds. The Sun will, therefore, in 7 hours, 28 minutes, 20·64 seconds, have moved eastwards such a distance, or arc, as the Moon would travel in 33 minutes, 32 seconds.

It will be remembered that the sidereal 19 years cycle occupies a slightly longer time than the synodical 19 years cycle, the difference being 18 minutes, 42·33 seconds. As the Moon's monthly course is eastwards, and as it is in conjunction with the Sun when it completes the synodical cycle, it follows that, when it

completes the sidereal cycle, it must be to the east of the
Sun by the distance represented by its progress during
18 minutes, 42·33 seconds, *less* the almost inappreciable
distance which the Sun itself will travel eastwards
during that time.

Now the Sun is not on the meridian of Greenwich
until 7 hours, 28 minutes, 20·64 seconds after the time
the Moon is in conjunction with it. During that period
both Sun and Moon will have travelled eastwards. As
the Moon makes its sidereal revolution in 27·32166 days,
the amount of its travel eastwards will be the proportion
which the time mentioned bears to the time of the
sidereal revolution, taking the latter as representing the
orbital circuit of 360 degrees. As the Sun makes its
apparent sidereal revolution in 365 days, 6 hours, 9
minutes, 9 seconds, the amount of its travel eastwards
will be the proportion which 7 hours, 28 minutes, 20·64
seconds bears thereto, taking the time of its apparent
sidereal revolution as representing 360 degrees. The
distance travelled by the Moon will therefore be 4° 6′
8·85″, and the distance travelled by the Sun 18′ 24·73″.

It follows that, when the Sun is next on the Greenwich meridian after the completion of the synodical cycle
of 19 years, the Moon will be 4° 6′ 8·85″, *less* 18′ 24·73″
(that is 3° 47′ 44·12″), to the east of it.

Let us now ascertain the positions of the Sun and the
Moon, when the former reaches the Greenwich meridian,
in relation to the point (or star) marking the completion
of the sidereal cycle.

Both the Sun and the Moon, at the time the synodical
cycle of 19 years is completed, are to the west of this
point by the distance which the Moon travels in 18

minutes, 42·33 seconds, which is about 10′ 16·17″. The
Sun and the Moon are, of course, in conjunction on the
completion of the synodical cycle. During the following
7 hours, 28 minutes, 20·64 seconds, the Sun, in its
apparent annual revolution, travels about 18′ 24·73″
towards the east, and, during the same time, the Moon
travels about 4° 6′ 8·85″ towards the east. In this
interval of 7 hours, 28 minutes, 20·64 seconds, the
meridian of Greenwich is, by means of the Earth's diurnal
rotation, turned directly towards the Sun, or, to put it
otherwise, the Sun is on the meridian at Greenwich 7
hours, 28 minutes, 20·64 seconds after being in conjunction with the Moon. Therefore, when this occurs
the Sun will be about 18′ 24·73″ to the east of the
position in the heavens which it occupied at the time the
synodical revolution was completed. Its position at that
time was, however, about 10′ 16·17″ to the west of the
point marking the completion of the sidereal cycle, so
that, when it is on the meridian of Greenwich it will be
about 8′ 8·56″ to the east of the same point. The Moon
has, during the time intervening between the completion
of the synodical cycle and the appearance of the Sun on
the meridian of Greenwich, moved eastwards by about
4° 6′ 8·85″. Its position when the synodical cycle was
completed was 10′ 16·17″, to the west of the point
marking the completion of the sidereal cycle. Therefore,
when the Sun is on the meridian of Greenwich, the Moon
will be 4° 6′ 8·85″ *minus* 10′ 16·17″ (that is 3° 55′ 52·68″)
to the east of the same point. It will, therefore, be
3° 47′ 44·12″ to the east of the Sun.

Thus, at the end of the 19 years cycle, on the first
ensuing return of the Sun to the meridian of Greenwich,

the relative positions in relation to the meridian are as follows, supposing the point of completion of the sidereal revolution of the Moon to be marked by a star.

Moon, 3° 47′ 44·12″ E.   *Sun*, on the meridian.   *Star*, 8′ 8·56″ W.

It follows that the star, the Sun, and the Moon, which, at the beginning of the 19 years cycle were all, as we have supposed, on the meridian of Greenwich, will occupy, at the end of the cycle, a stretch of 3° 55′ 52·68″, the Sun being on the meridian, and the Moon and the star being, respectively, to the east and to the west of the Sun. The distance separating the Sun and the Moon will, however, be about twenty-eight times the distance separating the Sun and the star. As the apparent diameter of both the Sun and the Moon is about half a degree, and as our measurements are applicable to the distances separating the centres of their disks, the *apparent* separation will be much less than we have found. The star will, in fact, be covered by the Sun, and the distance separating the edges of the disks of the Sun and the Moon will be only about 3¼ degrees. The separation is therefore small; but, still, the relative positions are manifestly different from those at the beginning of the 19 years period.

Let us now see whether the original relations of the meridian and the Sun and the Moon to each other are more accurately reproduced in a multiple of these 19 years cycles, than in the 19 years period itself. Our object is to get a period which will be as nearly as possible an exact number of days. We have seen that, in the 19 years cycle, the sidereal revolutions occupy 6939·70164 days, and the synodical revolutions 6939·68865 days. Although these revolutions occupy so

nearly the same time, the restoration of the original state
of matters is considerably departed from, before the same
meridian again faces the Sun.

It will at once be noticed that if we multiply these
periods by 3, we shall have very nearly an exact
number of days. Thus, three times 6939·70164 days are
20819·10492 days, and three times 6939·68865 days are
20819·06595 days. The fractions are quite insignificant.
This is the period applicable to 762 sidereal revolutions
of the Moon, and 705 synodical revolutions. The separation in time, between the completion of these series of
revolutions (·03897 of a day, which is equivalent to
56 minutes, 7 seconds), is three times the separation
occurring at the completion of the 19 years cycle of
revolutions. Let us see how this relates to the meridian
of the commencement of the cycle, in view of the three-
cycle period being completed so much nearer to the
meridian than the single cycle of 19 years.

As 705 synodical revolutions occupy 20819·06595
days we have the Sun and Moon in conjunction ·06595
of a day—that is, 1 hour, 34 minutes, 58 seconds—after
the Sun has passed the meridian. Let us now restore
the Sun to the meridian and see what would be its
position in relation to the Moon, and also to the star
which marks the completion of the sidereal period. As
the Moon, in its eastward course, does not come into
conjunction with the Sun until 1 hour, 34 minutes, 58
seconds after the Sun has crossed the meridian, it will,
when the Sun is on the meridian, be to the west of the
Sun. As the Sun, in its apparent annual revolution,
travels eastwards, the Sun also, will, when crossing the
meridian of Greenwich, be slightly to the west of the

position which it occupies 1 hour, 34 minutes, 58 seconds later. The distance which the Moon will travel in 1 hour, 34 minutes, 58 seconds is about 52′ 8″, and the distance, in its apparent annual eastward revolution, which the Sun will travel in the same time is 3′ 54″. If, therefore, we subtract 3′ 54″ from 52′ 8″, we shall get the distance which the Moon is west of the Sun, when the Sun is on the meridian of Greenwich for the last time before the completion of the 705th synodical revolution. The distance is therefore 48′ 14″.

Now, the sidereal revolution is not completed until ·10492 of a day after the Sun has passed the meridian—that is 2 hours, 31 minutes, 5 seconds—and, as the Moon has to travel eastwards to complete it, the star marking the completion of the revolution, must, when the Sun is on the meridian, be to the east of the Moon. The distance which the Moon will travel in 2 hours, 31 minutes, 5 seconds is about 1° 22′ 57″, so that, when the Sun is on the meridian, the Moon will be about that distance to the west of the star which marks the completion of the sidereal revolution. We have just seen that, at the same time, the Moon will be about 48′ 14″ to the west of the Sun. If, therefore, we subtract 48′ 14″ from 1° 22′ 57″, we shall have the distance which the Sun is west of the same star. The distance is consequently 34′ 43″. Thus, when the meridian faces the Sun, at the nearest approach to the completion of a cycle of three periods of 19 years each, the star, from which the sidereal cycle was reckoned, will be slightly to the east of the meridian—the distance being about 34′ 43″—and the Moon slightly to the west of the meridian—the distance being about 48′ 14″.

K

On the completion of 19 years, and three times 19
years, respectively, the position of the Sun, Moon, and
star, in relation to the meridian of Greenwich, is therefore
as follows :—

19 years period—
    Moon, 3° 47′ 44·12″ E.   *Sun*, on the meridian.   *Star*, 8′ 8·56″ W.
57 years period—
    Star, 34′ 43″ E.   *Sun*, on the meridian.   *Moon*, 48′ 14″ W.

It is evident that as regards these three main factors—
(1) the position of the Earth in its diurnal rotation, (2) the
position of the Sun and Moon in relation to each other,
and (3) the situation of the Sun and Moon in the heavens
—the state of matters prevailing at the commencement is
restored to a much greater extent in the 57 years period
than in the 19 years period.

Let us now notice briefly the position on the completion of the intervening period, which is 38 years—
being two cycles of 19 years each. In this period there
are 508 sidereal revolutions of the Moon, and 470
synodical revolutions, the time occupied by these, respectively, being 13879·40328 days, and 13879·37730 days.
The difference is thus ·02598 of a day, and the sidereal
period is the longer.

As ·37730 of a day is equivalent to 9 hours, 3 minutes,
18·72 seconds, it will be seen that the Sun will be very
far removed from the meridian of Greenwich before it
comes into conjunction with the Moon, on the completion
of the 38 years period. Since 1 hour represents 15 degrees of longitude in the diurnal rotation, 9 hours, 3
minutes, 18·72 seconds, must represent about 135° 49′ 41″,
which is therefore the distance west of the longitude of
Greenwich at which this synodical cycle is completed.

If, therefore, we restore the Sun to the meridian of Greenwich, we shall have to shift the Moon to the west of the Sun by the amount of its travel during 9 hours, 3 minutes, 18·72 seconds, *less* the distance apparently travelled by the Sun in the same time. The distance travelled by the Moon in the time mentioned is about 4° 58′ 17·2″. The Sun takes 365 days, 6 hours, 9 minutes, 9 seconds, to complete its apparent annual revolution of the heavens, so that, in 9 hours, 3 minutes, 18·72 seconds, it will, on the average, travel eastward to the extent of 22′ 18·7″. The distance separating the Sun and the Moon when the Sun is on the meridian of Greenwich will therefore be about 22′ 18·7″ less than 4° 58′ 17·2″—the actual separation being, consequently, about 4° 35′ 58·5″.

Now, as it takes longer to complete the sidereal cycle (13879·40328 days), than it does to complete the synodical cycle (13879·37730 days), the point marking the completion of the sidereal cycle must be to the east of the Moon even when the synodical cycle is finished, and the Sun and Moon are together. Therefore, when we remove the Moon to the west of the Sun—as we must do to obtain the relative positions at the time the Sun crosses the Greenwich meridian immediately before the 38 years cycle is finished—we remove the Moon still farther to the west of the point at which the sidereal cycle is finished. The time it takes the Moon to complete the sidereal cycle after completing the synodical cycle is ·02598 of a day, and, in that time, the Moon will travel eastward about 20′ 32·4″.

The separation between the Sun and the sidereal point, when the Sun is on the meridian at Greenwich, will be the distance separating them when the synodical revolution is completed—which is, as we have seen, 20′ 32·4″—*plus* the distance which the Sun has to travel eastwards, after crossing the Greenwich meridian, before the synodical cycle is completed. The latter is, as we have mentioned, about 22′ 18·7″, being the proportion of the annual eastward revolution of the Sun which occurs in 9 hours, 3 minutes, 18·72 seconds. The separation between the Sun and the sidereal point, at the time the Sun is on the Greenwich meridian, is, therefore, 42′ 51·1″. The Sun will then be to the west of the sidereal point, and the Moon, as we have found, still farther to the west.

If, now, we consider the point at which the sidereal cycle will be completed as marked by a star, the relative positions of the Sun, Moon, and star, in relation to the meridian, at the occurrence of noon at Greenwich immediately prior to the completion of the 38 years cycle, will be as follows:—

Star, 42′ 51·1″ E.   *Sun*, on the meridian.   *Moon*, 4° 35′ 58·5″ W.

We can now compare the three periods with which we have dealt, being 19 years, 38 years, and 57 years, respectively. The following is the comparison.

19 years period—
    Moon, 3° 47′ 44·12″ E. *Sun*, on the meridian. *Star*, 8′ 8·56″ W.
38 years period—
    Star, 42′ 51·1″ E. *Sun*, on the meridian. *Moon*, 4° 35′ 58·5″ W.
57 years period—
    Star, 34′ 43″ E. *Sun*, on the meridian. *Moon*, 48′ 14″ W.

The stretch of the three objects—Sun, Moon, and star —at the end of the different periods in their order, will be:—

        19 years period ...   . 3° 55′ 52·68″
        38 years period ...   . 5° 18′ 49·6″
        57 years period ...   . 1° 22′ 57″

It is evident that, in the longest of these three periods, the original relations of the Sun, the Moon, and the terrestrial meridian recur with considerably greater accuracy than in the two shorter periods. Indeed, if we allow for the diameters of the Sun and the Moon, it might be said that their apparent relations are almost exactly reproduced in 57 years.

Just as the 19 years period—which is, in fact, the period known as the "Metonic cycle"—appears to be the natural cycle of the Sun and Moon in their relations to each other, so the 57 years period appears to be the natural cycle of the Sun and Moon in relation to the Earth's diurnal rotation. This being so, let us see how the period fits in with the other classes of revolution of the Moon to which we have adverted, and also with the various movements of the Earth other than the diurnal rotation.

In the 57 years cycle there are, as we may remember, 762 sidereal revolutions of the Moon occupying 20819·10492 days, and 705 synodical revolutions occupying 20819·06595 days. We have to ascertain the relation the period of 20819 days bears to the anomalistic, the tropical, and the nodical revolutions of the Moon.

The anomalistic revolution occupies 27·55460 days. If, therefore, we divide 27·55460 days into 20819 days, we shall ascertain at what stage of its anomalistic revolution the Moon will be on the completion of the cycle, as compared with its position at the commencement of the cycle. We find that, in the period mentioned, there will be 755 complete anomalistic revolutions, and that the Moon will have been for 15·277 days engaged in the 756th anomalistic revolution, which it will still require

12·2776 days to complete. Thus, if we suppose the Moon to have been in perigee when the 57 years cycle began, it will be only slightly (1·4997 days), beyond apogee when it is completed ; and, if we suppose it to have been in apogee at the beginning, it will be a similar distance past perigee at the end. Thus, in two of the cycles it would be again in almost the original position. If it began at perigee, it would end 2·9994 days (say 3 days), beyond perigee. This would certainly have a modifying effect on the completeness of the 57 years cycle.

The tropical and nodical revolutions do not seem to be of any great importance to our subject. The tropical revolution occupies 27·32156 days, so that, in 20819 days, there will be almost exactly 762 tropical revolutions, the Moon requiring only ·02872 of a day to complete the 762nd revolution. The nodical revolution occupies 27·21222 days, so that, in 20819 days, there will be 765 complete nodical revolutions, and the Moon will have advanced 1·65170 days in the 766th revolution.

As regards the different classes of lunar revolution, the anomalistic is, therefore, the only one which shows a marked difference in its stage at the completion of the 57 years cycle as compared with the beginning, and it is a revolution of some importance to our subject.

Now, in regard to the orbital movements of the Earth, what is the state of matters after an interval of 20819 days as compared with their previous state ? This number of days is exactly equivalent to 57 years, subject to the exception arising through certain of the century years (those not exactly devisible by 400), not being leap-years. Thus, if we reckon 57 years from the beginning of 1850 to the end of 1906, we have 13 leap-years and

44 common years, the number of days being 20818,—one additional day being thus required for the cycle. If, however, we reckon 57 years so as not to include one of these century years, as, for instance, from the beginning of 1901 to the end of 1957, we have 14 leap-years and 43 common years, and the number of days is exactly 20819. Thus the completion of the period finds the Earth, as regards its tropical revolution, in practically the same position as at the commencement. The sidereal revolution of the Earth has already been dealt with by our consideration of the relative positions of the Sun, and the point of the heavens marking the completion of the Moon's sidereal revolution.

The only remaining revolution of the Earth is that which is accomplished in what is known as the anomalistic year. This year is of the same character as the anomalistic month, and thus marks the changes which occur in the positions of perihelion and aphelion in the Earth's orbit. These points revolve around the orbit in the same direction as the Earth itself, and thus the Earth has to accomplish rather more than a revolution between the recurring perihelions. The sidereal year, which is the true period of the Earth's revolution, is 365 days, 6 hours, 9 minutes, 9 seconds, while the anomalistic year is 365 days, 6 hours, 13 minutes, 48 seconds. The difference is thus 4 minutes, 39 seconds, which, in 57 years, would be 4 hours, 25 minutes, 3 seconds. The difference between the tropical year (365 days, 5 hours, 48 minutes, 46 seconds), and the anomalistic year is 25 minutes, 2 seconds, which would, in 57 years, amount to 23 hours, 46 minutes, 54 seconds. We cannot suppose that even the longer of these times—which is less than 24 hours—could have any appreciable effect in the way of modifying the weather. Whether the Earth is in perhelion, or 24 hours past perhelion, would seem to be a detail of utter insignificance in its atmospheric bearing.

One other matter which may have a slight bearing on the problem, and which has been already touched on, is the rotation of the Sun itself. Is it, approximately, the same side of the Sun which faces the Earth after 20819 days as faced it at the beginning of that period? There is some difficulty in answering this question, owing to the fact that the Sun does not rotate as a whole. At the solar equator, the period of rotation is 24 days, 2 hours, 11 minutes; while, at 30° N. latitude, it is 26 days, 9 hours, 46 minutes, and, at 30° S. latitude, 26 days, 12 hours, 50 minutes. The period is least at the equator, and increases with the solar latitude; but it is not the same for similar latitudes. If we take the three latitudes mentioned, then, for 20819 days, the result would be as follows :—

Solar Equator	864 rotations and	4 days.	9 hrs.	36 mins.
„ Latitude 30° N.	788 „	10 days.	7 hrs.	52 mins.
„ „ 30° S.	784 „	15 days.	18 hrs.	40 mins.

It is evident that, with the period of rotation varying continuously, correspondence cannot be attained throughout the body of the Sun. So far as can be judged, the solar meridians have no difference in effect on the terrestrial atmosphere; so that this matter is of no appreciable importance to our present study.

As regards solar disturbance and its possible effect on the weather, it would seem that, if the sun-spot period is correctly estimated at about 11½ years, there will be almost exactly five Sun-spot periods in 57 years, so that, on the expiry of 57 years, the Sun-spot period will be practically at the same stage as at the commencement.

The question of rotation does not, as we have seen, arise in connection with the Moon, as the Moon's period of rotation on its axis is identical with the period of its sidereal revolution; the same side of the Moon being thus always turned towards the Earth.

Local changes in the weather may possibly result from terrestrial causes, artificially produced, such as the growth, or the destruction, of forests; or the formation, or the removal, of lakes or swamps. They may also be occasioned by atmospheric disturbances arising from volcanic action. The effect on the great atmospheric ocean of such alterations on parts of the Earth's surface, or even of the disturbances arising from volcanic action, may be compared to the effect on the waters of the ocean of a stone dropped into the sea. Concentric ripples are sent out which, at a short distance, are lost in the movement of the waters. The weather changes resulting from such causes are, probably, purely local or of comparatively brief duration. Such causes of change do not seem to have any special bearing on our present enquiry.

We began by supposing that, as in the case of the waters of the ocean, the Sun and the Moon, in combination with the diurnal rotation, will cause tides and currents in the great ocean of the atmosphere, and that these tides and currents are the causes of the variations in the weather which we experience at the bottom of the atmospheric ocean, on the surface of the Earth. We conjectured that, if this were so, the recurrence of the same relation between the Sun and the Moon and the

Earth would cause a recurrence of exactly the same weather all over the Earth. We have seen that there is no exact recurrence of the relations between the Sun and the Moon and the Earth, in all their peculiarities, but that there are certain varying degrees of approximate recurrence. We have noticed that, in the course of the year, a certain approximation to a return of the relations occurs after 13 sidereal or 12 synodical revolutions of the Moon, the former occupying 355·18158 days, and the latter 354·36708 days. We have found that this approximation—even although the divergence is comparatively great—is characterized by the return of the same *type* of weather, as is evidenced by the recurring seasons. We have surmised that, if this happens when the recurrence of the relations is so evidently incomplete, a more close recurrence of the relations of the Sun, the Moon, and the Earth should be associated with a more marked repetition of the weather.

We have found that, on the expiry of 19 years, the Sun and the Moon return to very nearly their original relations to each other, but that the position of the Earth, in its diurnal rotation, is absolutely different from what it was at the commencement, and that, before the Earth regains its original position, the relations of the Sun and Moon have somewhat changed. Nevertheless, the state of matters at the end of 19 years, or rather 6940 days, is wonderfully similar to that at the commencement, and it would, therefore, be not improbable that the weather should be characterized by a corresponding similarity.

The true weather cycle, however, would seem to be 57 years, or, perhaps, we should say, 20819 days. On the expiry of that period the Sun and the Moon are in almost the same relation to each other and to the same meridian of the Earth as at the commencement; and, therefore, there appears to be strong reason to suppose that, on the expiry of that period, the weather should bear a close resemblance to the weather at the beginning. Thus, we may suppose that the 1st of January 1958 should have very similar weather to the 1st of January 1901; as, in the interval exactly 20819 days have elapsed. We have noticed, however, that, although the relations of the Sun and Moon to each other and to the terrestrial meridian are remarkably similar after 57 years to what they were at the beginning of the period, the actual *distance* of the Moon from the Earth is greatly altered. Thus, if the Moon were in perigee at the beginning of the period, it would be only a little past apogee at the end; if it were in apogee at the beginning it would be only a little past perigee at the end. As the Moon's distance in perigee is about 221,593 miles, and in apogee about 252,948 miles, we might have a difference in its distance of, perhaps, 30,000 miles, at the beginning of succeeding 57 years cycles.

It may be surmised that, with the apparent relations of the Sun, Moon, and Earth the same as at the commencement, this would not affect the similarity of the nature of weather, but merely its intensity. One high tide may differ from another high tide, but still both are high tides. Thus, if the Moon is nearer the Earth at the recommencement of the cycle than at its first commencement its power will be greater; if farther away, its power will be less; but, in both cases, its power will be exercised in the same manner. Of course, in the double cycle, the *distance* of the Moon is very similar to what it was at the beginning. The double cycle would be 114 years, but the approximation to an exact return of the relations of the Sun, Moon, and meridian would not, in other respects, for that period be quite so close as for the 57 years cycle.

While the *differences*, which arise in connection with the return of the Sun and Moon and the terrestrial meridian to similar relations after the lapse of 57 years, necessarily preclude the recurrence of an *exact* repetition of the weather, there yet appears to be strong reason to believe that a close similarity should occur, so that, in the true sense of the word, a cycle should exist. Nevertheless, in each succeeding cycle the discrepancies will become more evident, except only in so far as affected by the relative distance of the Moon in each double cycle as explained above. Subject to the allowance to be made for this peculiarity, it might be discovered by continued observation—in the event of the existence of the cycle being clearly established—in what *direction* the weather had changed, because, on the return of the same period once again in the following cycle, a further change in the same direction might reasonably be looked for. This applies to the changes brought about through the fact that the return of the Sun and the Moon and the terrestrial meridian to the former relations is not absolute, but merely approximate.

The weather has, to a great extent, persistently eluded the efforts which have been made for centuries to bring it under the dominion of known laws. The wind, even to this day, " bloweth where it listeth, and we hear the sound thereof but cannot tell whence it cometh and whither it goeth." Much progress has, however, undoubtedly been made in our knowledge of meteorology. If we could now prove with certainty the existence of a definite weather cycle we should make a great advance in the direction of bringing the weather itself under the dominion of known laws. We cannot doubt that the weather is as much subject to law as the Earth itself, as the Sun, the Moon, or the other planets. By slow degrees knowledge has been gained as to the laws governing these bodies. By slow degrees knowledge is being gained as to the laws governing the weather. By completely establishing the existence of a weather cycle, the course to be taken in the further investigation of the laws applicable to the weather, in all its variations, will be made more clear and intelligible—a result which is desirable, not only in the interest of science, but also, and perhaps more strongly, in the interest of humanity in its diverse pursuits, whether recreation, commerce, or industry.

THE SARGASSO SEA.

SYNOPSIS.

Experience of Columbus on his voyage of discovery
—Region called "Mar de Sargaço" by Columbus—
Extent of the Sargasso Sea—The ocean currents which
surround it—Experiments in ocean currents with
floating bottles—Information borne by ocean currents
from across the Atlantic before the discovery of
America—The "Gulf-weed"—The Gulf-weed fauna
—Tidal movement in Sargasso Sea—Depth of neigh-
bouring waters—Temperature and climate—Sargasso
Sea as "Dust-shoot" of the Atlantic—Erosion of land
by flowing streams—The Sargasso Sea and navigation
—The future of the Sargasso Sea—Probable effects of
the construction of the Panama Canal—Four methods
by which matter is accumulating in the bottom of the
Sargasso Sea—Growth of coral reefs and coral islands
—Possibility of the existence of under-currents in the
Sargasso Sea—Other similar regions—Conclusion.

THE SARGASSO SEA.

On the 16th of September, 1492, Columbus, when on his famous voyage of discovery, found his further progress threatened by great masses of sea-weed, which he compared to "extensive inundated meadows." His ships could scarcely make headway, and the sailors were in consternation. The discovery of the New World was hanging in the balance. Was it possible to go forward, or must he yield to the fears of the sailors and retreat? Happily, the iron nerve of the great discoverer gained the day. Slowly the ships continued on their course, and, in the result, the great Western Continent was added to the known world. This incident is an illustration of the fact, which so often comes under notice, that great events may depend upon very small causes.

The region in which Columbus passed through this embarrassing experience was called by him the "Mar de Sargaço,"—that is, the "Sea of Sea-weed,"—and the name has remained, the district being to this day known as "The Sargasso Sea."

This region of the globe is in many respects unique, and, as such, it forms a natural phenomenon of much interest.

The Sargasso Sea forms an extensive portion of the Atlantic Ocean. It may be considered as lying between

the parallels 20° and 35° N., and between the meridians 30° and 60° W. It extends in parts, however,
considerably farther west, perhaps indeed to the 75th
meridian. It is of an irregular elliptical shape, and its
size is given by Professor W. B. Carpenter, of the Royal
Institution, as being nearly equal to that of Continental
Europe, the extent of which is 2,740,100 square miles.
We may therefore take it that the area covered by the
Sargasso Sea exceeds two million square miles. Its
geographical position occupies the greater part of the
space lying between the north-eastern shores of the
West Indian Islands, and the north-western bulge of
the African Continent. It forms a vast expanse of still
water in the midst of the restless ocean.

Strange as it appears, it is yet the case that this
expanse of calm water is produced entirely by ocean
currents. The Sargasso Sea forms the centre of a
system of rolling streams. The most important of these
currents is the Gulf Stream, which comes sweeping out
of the Gulf of Mexico. It skirts the South of Florida,
where it is confined within narrow bounds, with a
velocity of about 80 miles a day, and flows north-eastward at the rate of about 48 miles a day. It forms the
northern boundary of the Sargasso Sea. As the Gulf
Stream continues its course it widens out, and its
rapidity further decreases. In the neighbourhood of the
Azores, which lie about 800 miles to the west of southern
Portugal, the border of this widened stream encounters
a southward current, which is known as the " North
African Current." A part of the waters which we have
traced from the Gulf of Mexico is drawn into this
current, and, united with the current into which it is

MAP SHOWING THE SARGASSO SEA

And the Ocean Currents by which it is surrounded.

Goll & Inglis, Edinburgh.

drawn, is turned southward, its continued course being no longer in the Gulf Stream, but in the North African Current. As this current flows southward, it impinges on the north-western shores of the African Continent. These shores in this part bulge out greatly into the Atlantic, extending farther westward as they are followed southward.

In consequence of this, and the changing temperature occasioned by the southward flow, the North African Current is deflected towards the west, so that the portion of its waters which was recently moving north-eastward in the Gulf Stream towards Europe is now, with the current with which it has united, flowing westward across the Atlantic. This westward current, as it approaches the northern coast of South America, encounters a current flowing north-westward from the equator, which is called the "Guiana Current," and which is simply an extension of the Main Equatorial Current. The combined waters have now a north-westward tendency, and flow into the Caribbean Sea, where they are deflected northward by the Isthmus of Panama and the lands of Central America. With increased rapidity, through the confined space which they now occupy by the intervention of the West Indian Islands, they flow into the Gulf of Mexico. A part of the water is thus doubtless once again brought back to the source from which it started. Once more it forms part of the current which flows around the Gulf of Mexico, and which passes into the Atlantic Ocean as the Gulf Stream. In the area between these converging currents lies the "Sea of Sargasso."

Some interesting experiments have been made in ocean currents with floating bottles. On 13th October, 1820,

Captain Parry, of H.M.S. *Hecla,* had a bottle cast overboard in latitude 56° 36′ N., and longitude 25° 45′ W.
The longitude is approximately that of the Azores, where,
as we have seen, part of the waters of the Gulf Stream
is drawn southward by the North African Current, but
the latitude is about 1,250 miles north of the Azores.
The bottle was picked up on the south coast of Iceland
(say, about latitude 63° 24′ N. and longitude 18° 54′ W.)
on 7th March, 1821. It had drifted about 470 miles
northward, during which it had passed eastward about
230 miles. In a direct line the distance travelled was
about 523 miles. If we assume that the bottle was
picked up on the same day as it was cast ashore, the
time occupied in the journey was 145 days, so that the
daily distance travelled—taking the course to have been
approximately a direct one—was on the average about
3·6 miles. This bottle was apparently carried along by
the northern fringe of the Gulf Stream,—which, as we
have seen, spreads out, fan-like, as it proceeds. On the
16th June, 1819, a bottle was thrown overboard from the
Hecla in latitude 53° 13′ N., and longitude 46° 55′ W.—
say, about 450 miles north-east of Newfoundland. On
the 29th of July, 1821,—that is 2 years and 43 days
after it was thrown into the sea,—it was picked up on
the south-east shore of Teneriffe, the largest of the
Canary Islands, at about latitude 28° 30′ N., and longitude 16° 30′ W. Supposing this bottle had travelled in
a direct course, the distance passed over would have
been more than 2,325 miles, which would give an average
rate of travel of 3 miles a day,—assuming that it was
picked up on the same day as it was cast ashore. It is
clear, however, that it did not travel in a course

approximating, to any extent, to the direct. It had evidently been carried by the current which sweeps southward past Labrador and the east coast of Newfoundland, into the Gulf Stream, the strength of the current carrying it so far into the stream as to escape the northern portion which spreads out towards northwestern Europe. It was afterwards carried by the Gulf Stream past the Azores, and was then conveyed by the North African Current to the spot at which it was picked up. It, in fact, passed along the margin of the Sargasso Sea for about one-third of its extent. In all probability it had been tossed backwards and forwards by opposing currents for a long time. It is, of course, impossible to say how long it may have been lying on the shore, before it was at last noticed and picked up, more than two years after starting on its adventurous voyage.

It has been estimated, from careful observation and calculation, that a floating object would in 13 months be carried by the sea-currents from the Canary Islands to Caracas on the coast of Venezuela, and that in 10 months it would accomplish the tour of the Gulf of Mexico, and arrive at the western shore of Florida. In from 40 to 50 days it would be swept by the full force of the Gulf Stream from the Straits of Florida to the coast of Newfoundland.

It was, indeed, through information borne by these ocean currents that Columbus was led to undertake his voyage across the unknown seas. Two corpses, which were described as "very broad-faced and differing in aspect from Christians," were washed up at Flores, one the islands of the Azores. A piece of wood, "strangely

wrought," was picked up on the shore to the west of Cape
St. Vincent, the south-western extremity of Portugal,
after a prolonged westerly gale.   Pieces of bamboo, also,
of an extraordinary size, cast ashore by the western
currents, were found on the beach at Porto Santo, one
of the Madeira Islands, by no less a person than the
Governor of the Island, Pedro Correa, a brother-in-law
of Columbus.   These repeated messages from the sea
had certainly no little influence in deciding the great
discoverer to brave the perils of the deep.

By these ocean currents, then, the Sargasso Sea is surrounded, while, in itself, it forms a region of undisturbed
calm.

The sea-weed, from which the Sargasso Sea takes its
name is of a different species from the forms of sea-weed
which occur on the British coasts.   It is called the
Sargassum bacciferum, which may be literally translated as "the sea-weed bearing berries."   It is commonly
known as the "Gulf-weed," from the supposition that it
was merely carried from the Gulf of Mexico and cast
into the Sargasso Sea, a supposition which has been
found to be erroneous.

The *Challenger* expedition passed a considerable time
in this part of the Atlantic in 1873, and made almost the
circuit of the Sargasso Sea.   Much attention was given
to the examination of the sea-weed which had given so
much trouble to Columbus on *his* expedition more than
four hundred years previously.   The late Professor Sir
C. Wyville Thomson, the Director of the civilian
scientific staff of the *Challenger*, describes the weed as
it appeared on the margin of the sea.   It was found to
consist of a single layer of feathery bunches, not matted

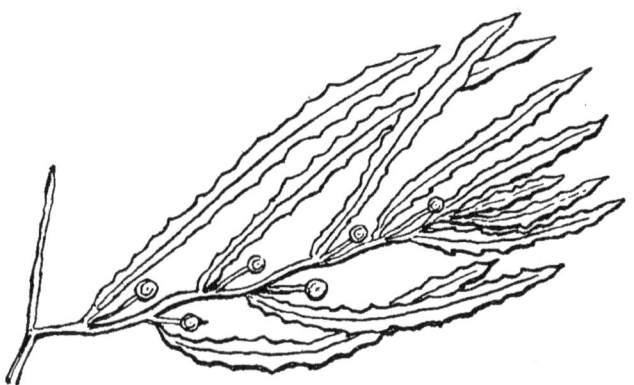

The Gulf Weed *(Sargassum Bacciferum).*—p. 152.

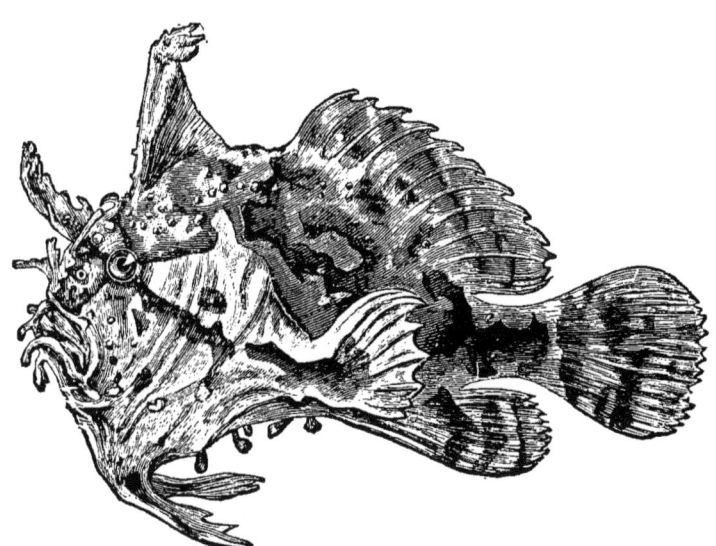

Antennarius Marmoratus. Natural size.—p. 154.
(One of the fauna of the Gulf Weed).

but floating nearly free of one another, only sufficiently entangled for the mass to keep together. Each tuft had a central, brown, thread-like branching stem studded with round air-vesicles on short stalks. Most of these air-vesicles, or air-cells, near the centre of the plant, were seen to be dead, and they were coated with a beautiful netted white polyzoon. It appears to be through the support afforded by the air-cells that the weed floats on the surface. After a time the encrusted air-cells break off, and, where there is much of the Gulf-weed, numerous little white balls are seen floating freely. It is from these air-cells, with their grape-like appearance, that the weed derives its specific name of the berry-bearing sea-weed—the *Sargassum bacciferum*. A short way from the central stem of the weed, willow-like leaves occur on the branches. The leaves are serrated, and those nearest the stem are brown and rigid, but farther away they are paler, fresher, and more active in their vitality. The general colour of the mass of the weed is olive in all its shades, but the golden olive of the young and growing branches much predominates. The colour is, however, greatly broken up by the delicate branching of the weed, blotched by the vivid white encrusting polyzoon, and riddled by reflections from the bright blue water gleaming through the spaces in the net-work. The general effect of the profusion of the weed in contrast with the blueness of the water is most pleasing.

Sir Wyville Thomson describes the Gulf-weed as forming floating islands, which, at the margin of the Sargasso Sea, are, usually, from a couple of feet to two or three yards in diameter. Some, however, are much larger.   He mentions having, on different occasions, come across fields several acres in extent, and he concluded that such expanses were probably more frequent near the centre of the area of the distribution of the weed.   Thus, he did not suppose that, even in the central portion, the weed covers areas many miles in circumference in absolutely unbroken continuity, but, on the contrary, that, here and there, openings of comparatively clear water must intervene.   In fact, while the separated accumulations at the margin appear to form small islands, the accumulations in the interior might probably be compared to large islands.

Sir Wyville Thomson found that the weed had inhabitants peculiar to itself, and he says that he knows of no more perfect example of protective resemblance than that which is shown by the Gulf-weed fauna. Animals drifting about on the surface of the sea, with such scanty cover as the single broken layer of the seaweed, must be exposed to exceptional danger from the sharp-eyed sea-birds hovering above them, and from the hungry fishes searching for prey beneath; but it was found that these creatures, one and all, imitate in an extraordinary way, both in form and colouring, their floating place of abode, and consequently, one another, so that both birds and fishes are readily deceived.

Amongst these inhabitants of the weed was found a grotesque little fish (*antennarius marmoratus*) resembling the English "fishing frog" (*lophius piscatorius*) which is often thrown up on the British Coasts, and is conspicuous for the disproportionate size of its head and jaws, and for its general ugliness and apparent rapacity. Although many specimens of this little fish were examined by the scientific staff of the *Challenger*, not one was found to exceed 50 millimetres—that is, about two inches—in length. It was uncertain, however, whether the specimens found had attained their full size. This little fish has the peculiar habit of constructing a nest of the Gulf-weed. It binds the weed into a bundle in the form of a compact ball, a little larger than a cricket ball. The ball is bound together by very strong transparent strings formed of a viscid secretion from the fish, and the spaces among the leaves, or fronds, of the weed are filled with the eggs. A number of these balls were found by the *Challenger* floating on the margin of the Sargasso Sea, and were picked up and examined.

Another frequent inhabitant of the Gulf-weed is the shell-less mollusc called *scillæa pelagica*, which may be freely interpreted as "the bulb-like surface jelly-fish." Swarms also were seen of a little short-tailed crab which is known as the *nautilograpsus minutus*. It abounded, not only on the weed, but on every floating object, and it was noticed to vary in colour with whatever it happened to inhabit. Sir John Murray, who was a member of the scientific staff of the *Challenger*, found that these little crabs had the most lively appreciation of their own proper place of abode. He amused himself by dislodging some of them, but they always made the most vigorous efforts to regain their original position. It was noticed that, although every floating thing was full of them, they were rarely seen swimming free from their floating support.

Sir Wyville Thomson found that the Gulf-weed animals, fishes, mollusca, and crabs did not simply imitate the colours of the weed which they inhabited. This would have produced suspicious patches of continuous olive. They were found to be, like the weed itself, blotched over with bright opaque white, the blotches being sometimes irregular, but generally rounded. At a little distance these white blotches were absolutely indistinguishable from the patches of polyzoa on the weed, to which we have referred.

The Gulf-weed was also found to be inhabited by numerous other creatures of similar type to those mentioned. In fact the Sargasso Sea is the abode of countless myriads of small crustaceans and minute organisms.

The tidal rise and fall has no appreciable effect on the weed of the Sargasso Sea. It might be supposed that the tidal wave in its forward movement would clear away the surface matter, and carry it onward to be deposited on far distant shores, just as each tide leaves, at high-water mark, a row of sea-weed along our coasts. This however is not the case. As a matter of fact, in the Atlantic, the rise and fall of the tide is comparatively slight, not exceeding 10 or 12 feet in the open, or deep, sea. The effect of the tide on the Sargasso Sea is also probably impeded by the surrounding currents acting as protecting barriers. Thus the tidal wave will, as regards surface agitation, result in little more than a mere undulatory movement.

The average depth found by the *Challenger* near the margin of the Sargasso Sea was about 2750 fathoms, say 16,500 feet. The trawl showed the bottom to consist of grey mud, and brought up many small shell-fish.

The temperature in the neighbourhood of the Sargasso Sea was found to vary from about 23·3° C. (say 73·94° F.) at the surface, to 1·9° C. (say 35·42° F.) at the bottom.

This was in the month of March. Of course, different temperatures were obtained at different parts and in different seasons, but the variation from those mentioned was not great.

Dr. Alexander Buchan states that regions of light and variable winds and calms occur at the higher limits of the north and south trade winds. Of these regions of calms, the most important, in his opinion, is that marked off by the high pressure in the North Atlantic, between the United States and Africa. "This is," he says, "the region of the Sargasso Sea, where the weather is characterized by calms and variable winds, and the ocean by its comparatively still waters." These parts are known to seamen as the "horse latitudes," the name having, it is supposed, originated through the old navigators having in these districts, through the tedious calms, frequently thrown overboard the horses which they were transporting to America and the West Indies. The "horse latitudes" are essentially different from the equatorial region of calms. The latter is a region of low atmospheric pressure at the meeting of the north and south trade winds, where the climate is distinguished by its general sunlessness and heavy rainfall. On the other hand, the calm region of the Sargasso Sea is situated mainly above the tropics, and has an atmospheric pressure abnormally high, with clear skies, and, generally, sunny and bright weather, with occasional squalls. Columbus describes the weather which he experienced in the Sargasso Sea, when on his great voyage of discovery, as being characterized by "most temperate breezes, the sweetness of the mornings being most delightful, the weather like an Andalusian April, and only the song of the nightingale wanting," and this at the time that his men were murmuring and complaining in their alarm at the thick matting of sea-weed which retarded their progress.

The supposition that the weeds of the Sargasso Sea were not indigenous, but were carried by the Gulf Stream from the Gulf of Mexico, was disproved last century through the investigations of Captain Laps, a Frenchman. He found, beyond doubt, that the weed is produced where found, the growth taking place on the surface of the water. The old plants, which have lost their leaves, send out young shoots, and these develop into new plants.

Not only, however, is the Sargasso Sea the habitat of these masses of sea-weed and these myriads of small organisms, it is also what we may call the "dust-shoot" of the Atlantic. Into this region, according to Professor Carpenter, is collected a large proportion of the drift or wreckage which floats about the North Atlantic. In this huge receptacle, it is stated that at least two-thirds of all the matter carried along by the Gulf Stream ultimately finds its resting place. Here are to be seen "huge trunks of trees, torn from the forests of Brazil by the waters of the Amazon, and floated down far out to sea, until they were caught and swept along by the current; logwood from Honduras; orange trees from Florida; canoes and boats from the Islands, staved in, broken, and bottom upwards; wrecks, and remains of all sorts, gathered from the rich harvest of the Atlantic; whole keels, or skeletons of ruined ships, so covered with barnacles, shells, and weed, that the original outline is entirely lost to view; and here and there a derelict ship transformed from a floating terror of the deep into a mystery put out of reach of man in a museum of unexplained enigmas."

The earthy matter, carried along by the different currents, which ultimately finds its way to the bottom of the Sargasso Sea, must, in itself, be considerable. Sir Archibald Geikie, in his geological researches, gave much attention to the erosion of land by flowing streams, and his conclusions are most interesting. He says :—

"The true gauge of the present yearly waste of the surface of the land is furnished by the amount of mineral matter carried every year into the sea by rivers. This mineral matter is partly in mechanical suspension, partly in chemical solution, and is, to no small extent, pushed in the form of shingle and sand along the bottoms of the streams. . . . If we take the ratios furnished by the Mississippi as a fair average, which, from the vast area and varied climatal and geographical characters of the region drained by that river, they probably are, then we learn that $\frac{1}{6000}$th of a foot is worn away from the general surface of the land every year. At this rate, if the present erosion could be sustained, the whole American Continent, of which, according to Humboldt, the mean height is 748 feet, would be worn down to the sea-level in about $4\frac{1}{2}$ millions of years—a comparatively short time in geological chronology."

Of course, it is probable that only a relatively small part of the solid matter torn away by the force of rivers and tides and carried into the ocean currents which bound the Sargasso Sea will be driven by these currents into that portion of the Atlantic, but although this is so, the amount which, in the course of the centuries, will be so collected must in the aggregate be enormous. The mode of the formation of the Sargasso Sea must neces-

sarily lead to the currents driving such solid matter into
the protected area rather than drawing it away therefrom,
assuming that, to any extent, the movement of the water
at the bottom harmonizes with that at the surface.

The peculiar character of the Sargasso Sea has been
neatly illustrated by comparing it to a basin half-full of
water with chips of wood and light particles floating in
it. If a circular motion is given to the water around the
margin, while the central part is left undisturbed, the
floating matter is gradually collected into the calm water
at the centre. In the same way the floating bubbles in a
tea cup gather together in the centre when the tea is
stirred.

Although the Sargasso Sea lies geographically between
countries having extensive shipping communications with
each other, it is unnecessary for the main shipping routes
to pass through its denser parts. The Sargasso Sea is, in
fact, so placed that the hindrance which such a profusion
of sea-weed would offer to shipping is, to a great extent,
obviated. It may be said to form a triangle, the sides of
which form the principal ocean routes, so that, like the
currents, these routes form its boundaries. Of course,
however, in certain routes, the bounds of the Sargasso
Sea may necessarily, to some extent, have to be encroached
on. Columbus, on his first voyage to America, appears
to have passed through almost the whole length of the
weed-covered district; but, probably, this has never been
done since then. With the development of navigation,
advantage is, as far as possible, taken of the currents for
assisting the vessel in its course, so that an area of continued stagnation is avoided. Although the district
covered by the Gulf-weed is not absolutely continuous, it

is recognized that, both to sailing vessels and steamships, it would be a matter of no small danger to venture far into it. A recent writer on this subject expresses a doubt whether a sailing vessel would be able to cut her way into the thick net-work of weed, even with a strong wind behind her, and, he considers, that, if she succeeded in doing so, "there would be almost certain danger that in the calm regions of the Sargasso Sea the wind would suddenly fail her altogether leaving her locked helplessly amid the weed and the drift and wreckage, without hope of succour or escape." The same writer expresses the opinion that no prudent steamship captain is ever likely to make such an attempt, "for it would certainly not be long before the tangling weed would altogether choke up his screw and render it useless."

The Sargasso Sea is, in reality, neither land nor water. It has been described as resembling meadows in the sea, and, in this complex character, it is unsuited for any means of locomotion exclusively adapted for either sea or land. Notwithstanding the enterprise of explorers, a district like this seems singularly well protected from complete exploration. It might be recommended to any person who wishes to go into retreat and to escape the restrictions of society,—to any person desiring to lead a hermit existence free from all risk of intruders. He would find the climate as nearly perfect as possible. According to Columbus "the nightingales alone are wanting." No doubt, however, there would be difficulties in the way of obtaining supplies, and the taking of exercise would necessarily be confined within very narrow limits!

The future of the Sargasso Sea is a question of much

interest. It has been suggested that, in course of time, the increasing pressure will result in solidifying the whole mass into coal for the supply of mankind in the distant ages. This is, of course, a mere conjecture. It seems most probable that any such extreme pressure cannot occur on the surface, whatever may happen at the bottom.

The opinion may be hazarded that, possibly, in the Sargasso Sea we have a small continent or, at least, an archipelago in the course of formation. We know that, for many centuries, this region has been the abode of continual vegetable growth. In such a situation growth is extremely rapid and profuse. It is impossible to suppose that the identical plants through which the ships of Columbus cleaved their way are those which now flourish on the surface of the sea. It seems most likely that, year after year, season after season, there is a constant succession of vegetable life, that the individual plants have their beginning, their brief existence, and their disappearance. We may reasonably surmise that as each plant in succession decays its buoyancy will disappear and its tendency will be to sink to the bottom. For a time support is afforded by the neighbouring growth, but, by degrees, the dead plants outweigh their support, and then a portion of the growth falls to the bottom and a portion of the surface is temporarily left clear. If, in these circumstances, the surface of the sea continues practically covered with the weed, what must be the condition of matters at the bottom? The conclusion seems irresistible that, by slow degrees, extending through ages, the Sargasso Sea will, by the continuous accumulation of the vegetable growth, and of the matter deposited by the surrounding currents, be completely filled up.

This process must be, to some extent, hastened by the masses of small organisms, including shell-fish and crustaceans, which pass their short existence in the calm area of the Sea, and which are unsuited for the rougher waters of the ocean. These, like the vegetable matter, come into existence, develop, and perish, and, in most cases, then sink to the bottom of the still waters of this extraordinary Sea.

In this connection it may be noticed that it is not at all impossible that the construction of the Panama Canal, if the Canal is formed at sea-level—that is to say without locks—may have some effect on the Gulf Stream, and, thereby, on the Sargasso Sea. It would seem not improbable that, in the circumstances mentioned, a portion of the Gulf Stream will be deflected into the Canal, so as to flow through it from the Atlantic to the Pacific, instead of continuing its passage around the Caribbean Sea and Gulf of Mexico. Such a deflection would neccssarily have a modifying tendency on the strength of the remaining part of the Stream. There can, however, be little doubt that any deflection so caused would be very insignificant in amount in comparison with the extent of the Gulf Stream. It would probably, indeed, be inappreciable in its effect on the Gulf Stream, and consequent effect on the Sargasso Sea.

It would appear that, by four separate methods, matter is slowly accumulating at the bottom of the Sargasso Sea. These are (1) the decay and fall of vegetable surface growth; (2) the perishing and sinking of myriads of small living creatures; (3) the drift of the flotsam

M

and jetsam of the Atlantic; and (4) the under-drift of non-buoyant matter.

We may take it that, to a certain extent, these substances—particularly the animal matter—will not remain permanently at the bottom of the sea; but, even if we make every allowance for possible displacement by living agency, it seems evident, that, with every passing year, the Sargasso Sea will, to a certain extent, be filled up.

We have, in the coral reefs and coral islands of the Pacific and Indian Oceans and other parts of the globe, illustrations of the vast growth which minute accretions will in time effect when innumerable agents are at work in their accumulation. The coral-producing zoophytes, which form these reefs and islands, are individually of small account. In comparison with the works which they produce, they are utterly insignificant. Their numbers, however, are uncountable, and, by their united perseverance, islands, which have subsequently become the abode of men, and reefs of gigantic dimensions, have slowly been built up. As examples of such islands, we have the Bermudas in the West Indies. These, according to Dr. G. C. Bourne, of Oxford, are entirely of coral formation to begin with—the higher lands being subsequently formed by sand accumulated by the wind, and gradually cemented into rock. As examples of reefs, we have the succession of reefs which which form the Great Barrier Reef on the north-east coast of Australia, which, according to the late Professor H. A. Nicholson, extend, with occasional breaches, for a distance of over 1,000 miles. These reefs are at an average distance from the shore of between 20 miles and 30 miles, the depth of the water on the inner side varying from 60 to 360 feet, and on the outer side ranging as high as 2,000 feet. These facts well illustrate the enormous increase which, in the course of ages, must occur in the accumulation of matter at the bottom of the Sargasso Sea, if the matter deposited remains, to a considerable extent, undisturbed.

This raises the question whether there is any likelihood of there being under-currents in the Sargasso Sea itself, whereby such accumulations may be prevented. This is a difficult question, and the only possible answer is that there is nothing, so far as has been ascertained, which indicates the existence of such currents, or at least, of currents of sufficient strength to remove the matter deposited, or to prevent its continued increase. Sir Wyville Thomson expresses the opinion that the movements of masses of underlying water are so slow, that, even if we had some feasible method of observation, the indications of movement, within a limited period, would be too slight to be measured with any degree of accuracy.

We have noticed that the *Challenger* expedition found the water in the neighbourhood of the Sargasso Sea to be very deep—about 16,500 feet. This, however, gives no indication of the depth of water in the interior of the weed-covered area. The Sargasso Sea may be considered as a very large lake in the middle of the ocean, its shores being bounded by ocean currents instead of by land. If the height of the adjoining land can be regarded as affording any information at all as to the depth of an inland lake, it is inversely rather than directly; and this can scarcely be supposed to apply to the circumstances of the Sargasso Sea. We have therefore no reliable information as to the present depth of this ocean lake.

Although the Sargasso Sea is the largest portion of the globe which bears the strange character of being, in the proper sense, neither sea nor land, it is not the only portion. There are in the Pacific Ocean, two similar districts of comparatively small size. Sir John Murray makes reference to these districts in the following passage.

"The general direction of surface circulation in the Pacific may be remembered by supposing the ocean divided into a northern and southern half by the Equatorial Counter Current. In the northern half, the water circulates in the direction of the hands of a watch, i.e., it passes up the west coast and down the east; while, in the southern half, the rotation is in the opposite direction—down the west coast and up the east; but the latter half does not exhibit the complete cycle so distinctly as the former. The centre of each area of circulation is occupied by a small Sargasso Sea, the northern being the more clearly defined; but neither approaches the well-known Sargasso Sea of the North Atlantic, either in definiteness, extent, or amount of weed."

The surface circulation of the Pacific has been found to be, on the whole, less active than that of the Atlantic, and it is supposed to be, in general, confined to levels very near the surface. The average depth of the Pacific is greater than that of the Atlantic, there being, indeed, in the Pacific Ocean, areas of greater depth than have been found in any other part of the Earth. The argument, therefore, which we have applied to the Sargasso Sea, properly so-called, would be very greatly weakened in its application to these weed-infested areas in the Pacific. The Gulf-weed is found not only in the regions to which we have referred in the Atlantic and Pacific Oceans, but, also, in the Indian Ocean and the Red Sea; but nowhere else does it exist in such extraordinary profusion as in the so-called Sea in the Atlantic Ocean, which has derived its name from it.

It would appear, from the facts which we have detailed, that the Sargasso Sea is at present in a state of transition, that what is to-day called the "sea" will—failing any important change in the enclosing currents—by and by be solid land. It is, no doubt, premature to suggest that the Powers, who are now on the look out for Colonial possessions and who are finding themselves forestalled in every direction, might keep an eye on the Sargasso Sea. Ages may still have to elapse before any portion of this area acquires sufficient solidity to support a national flag. But, although this is so, the whole facts appear to indicate, almost beyond doubt, that the time will come when the Sargasso Sea shall disappear from our maps, and, in its place will be shown a new tract of land or perhaps a new group of islands. Just as surely as the constant dripping wears out the rock, the continuous accumulation of solid matter in the Sargasso Sea must ultimately result in the extinction of the Sargasso Sea itself. But *when* shall the time come for the completion of the change? Who can say?

THE ZODIACAL LIGHT,
THE ZODIACAL BAND, AND
THE GEGENSCHEIN.

SYNOPSIS.

The different phenomena described—Ancient observation of the Zodiacal Light—Views of Kepler and other astronomers of past days—Modern investigations —Observations of the Zodiacal Band—Discovery of the Gegenschein by various observers independently— Visibility of the different phenomena—Explanatory theories and existing uncertainty — Meteors and meteorites in the solar system — Meteorites encountered by the Earth and the terrestrial atmosphere— Reflection of sunlight by meteoric bodies—Relative density of such bodies in line of sight at different hours, and their relative reflection of light earthwards—Effects of the eccentricity of the orbits of meteoric bodies— Possible explanations of Earth's shadow not being visible in the Gegenschein—Interception of sunlight by meteoric bodies—Observation of a Zodiacal Light in the neighbourhood of the Moon—Conclusion.

THE ZODIACAL LIGHT,
THE ZODIACAL BAND, AND
THE GEGENSCHEIN.

THE Zodiacal Light may be defined as a luminous tract of a conical or lenticular shape, visible occasionally in temperate latitudes, and generally in tropical latitudes, after sunset or before sunrise, extending upwards from the position of the Sun, and lying nearly in the direction of the ecliptic, its base being on the horizon and its apex at varying altitudes. It follows the setting sun and precedes the rising Sun. It must be distinguished from the rosy hue of the sky at and near the horizon, associated with sunset and sunrise. The Zodiacal Light has been frequently supposed to have a reddish-yellow tint, but, when it is seen under favourable conditions it is found to be distinctly white—its light being similar to that of the Milky Way. Any colour appearing in the Zodiacal Light is therefore of atmospheric origin.

The Light is brightest and of the greatest breadth near the horizon, its breadth in that position being from 20 degrees to 30 degrees, possibly even more. The breadth lessens with the distance from the horizon, or, we may say, the distance from the Sun, because when the Light has been perceived the Sun has invariably been below the horizon. Possibly, the Light might be distinguished during a total eclipse of the Sun, at whatever altitude, although there does not appear to be any record of it having been so observed. Probably, indeed, the light of the solar corona, which, even in a total eclipse of the Sun, extends beyond the lunar disk, would be sufficient to prevent the faint luminosity of the Zodiacal Light being visible. If the Zodiacal Light were distinguished during a solar eclipse, no doubt the brightest and broadest portions of the area of light would be the parts nearest to the Sun, east and west of the Sun, in the ecliptic. The Light, when observed after sunset or before sunrise, is seen to extend from the horizon to a distance of, perhaps—under favourable circumstances—110 degrees or 120 degrees from the Sun, possibly even more; the summit or apex occasionally appearing to be somewhat pointed, but generally to be rounded off.

The Zodiacal Band appears like a comparatively narrow extension of the Zodiacal Light, stretching from the highest part of either the eastern or western cone. It is of fairly uniform breadth, and reaches at times from the summit of the cone to the horizon opposite to that at which the Zodiacal Light appears; or from the summit of one cone to that of the other, when the cone of Zodiacal Light is coming into view above one horizon before disappearing from view below the opposite horizon, as sometimes occurs when the Light is exceptionally bright and the situation good. When the Zodiacal Band is thus perceived it presents to view a continuation of the Zodiacal Light around the whole arch of the heavens.

The Gegenschein is one of the most mysterious of natural phenomena. In fact, it is with difficulty astronomers have come to accept its existence. It is a faint light, and is usually circular, but sometimes apparently oval-shaped. It is much larger in apparent diameter than the Sun, and occupies a position in the heavens exactly, or almost exactly, opposite to that occupied by the Sun. As the Sun is on the meridian at mid-day, the Gegenschein is therefore on the meridian at midnight. The apparent diameter of the Sun is about half a degree. The apparent diameter of the Gegenschein is, owing to the faintness of the light, somewhat uncertain, and varies according to the brightness of the light. It is sometimes 10 degrees, sometimes as much as 20 degrees. The Gegenschein is a huge ghastly image of the Sun circling around the Earth at the greatest time-interval from the Sun. As the Sun rises in the east the Gegenschein sets in the west, as the Sun sets in the west the Gegenschein rises in the east.

The observations of the Zodiacal Light, although, from an astronomical point of view, but of yesterday, are yet ancient in comparison with those of the Zodiacal Band and the Gegenschein.

The Zodiacal Light is said to have been seen by the ancient Egyptians, who are supposed to have worshipped it. The Rev. Edmund Ledger, Gresham Lecturer on Astronomy, suggests that it is identical with what Pliny calls "*trabes*" or "the beam"; and that a reference to it may be found in the writings of Omar Khayyam, the Persian astronomer and poet, who, 800 years ago, used the words, "before the phantom of the *False Dawn* died."

In the Koran it is referred to as the "False Aurora," or "False Dawn"—the Mohammedans attaching importance to the accurate determination of the time of daybreak, in view of the commencement at that hour of the daily fast observed during the ninth month of the Mohammedan year.

In Europe, Kepler observed the Zodiacal Light apparently early in the 17th century. He concluded that it was the atmosphere of the Sun. Descartes, the French philosopher, wrote about it in 1630; and Childrey, in 1659, in his *Britannica Baconica* explicitly refers to it. The passage is quoted by the Rev. Edmund Ledger.

"There is another thing which I recommend to the observation of mathematical men, which is, that in February, and for a little before, and a little after that month (as I have observed several years together), about six in the evening when the twilight has almost deserted the horizon, you shall see a plainly descernible way (*i.e.*, track) of the twilight, striking up towards the Pleiades, and seeming almost to touch them. It is so observed any clear night. There is no such way to be observed at any other time of the year (that I can perceive) nor any other way at that time to be perceived darting up elsewhere.

It was not, however, until 1683, that the attention of astronomers was first prominently called to the Light. This was done by Dominic Cassini, Director of the Observatory at Paris. He observed the Zodiacal Light for the first time on 18th March 1683, and referred to it as "the Light which appeared in the Zodiac"—hence the name it still bears. He made a special study of it, and conjectured that it consisted of a flat luminous ring encircling the Sun nearly in the plane of the solar equator.

Laplace, one of the most celebrated mathematicians and astronomers of the early years of the 19th century, showed that the Zodiacal Light could not possibly be the solar atmosphere, because no atmosphere could rotate with the Sun if it extended beyond the distance where centrifugal force is balanced by gravity—a distance far within the observed extension of the Zodiacal Light from the Sun. He also showed that, even supposing the solar atmosphere could extend outwards from the body of the Sun to the observed distance of the Zodiacal Light, its appearance would be different from the conical or lenticular appearance of the Zodiacal Light, as the polar axis of the Sun, including the solar atmosphere, would be at least two-thirds of the equatorial axis. Laplace's investigations were therefore destructive of the conclusions of Kepler.

Since the days of Laplace, many observers have made a study of the Zodiacal Light. The late Professor Piazzi Smyth, Astronomer Royal of Scotland, and Professor A. W. Wright, of Yale, found by spectroscopic examination that the Light was almost exactly similar to that of faint diffused sunlight, such as is got in the last traces of twilight. The Rev. George Jones, of the U.S. frigate *Mississippi*, made most careful observations of the Zodiacal Light, during an expedition to and from Japan in the years 1853-5. He repeatedly observed the Light stretching upwards from both eastern and western horizons simultaneously. He thought, as a result of his investigations, that the Light must be occasioned by a ring of nebulous matter, not around the Sun but surrounding the Earth, within the orbit of the Moon. This supposition has not, however, been found to furnish a satisfactory explanation of the observed phenomena. Professor C. Michie Smith, of the Madras Solar Physics Observatory, has also given much attention to the Zodiacal

Light. He points out that the explanation suggested by the Rev. George Jones is untenable. "The best observations leave no room for the parallactic displacement which would be observed if there were such a ring round the Earth, and the absence of a large part of the luminous circle on ordinary occasions would be inexplicable on any such hypothesis." Professor Smith is of opinion that "we must look for the cause of the light to the existence of a mass of small bodies moving in orbits round the Sun," and that "the light is chiefly, if not entirely, reflected sunlight." The weight of modern astronomical opinion has supported this view, although the conclusion is still so uncertain that Professor Barnard, of the Yerkes Observatory, so recently as January, 1905, refers to the Zodiacal Light as a mystery still awaiting explanation.

The observations of the Zodiacal Band are all very recent. It appears to have been discovered, or first observed with some definiteness, only about the beginning of the eighth decade of the 19th century—say about the year 1871. Its outline is so uncertain that doubt has existed as to its identification as an extension of the Zodiacal cone. Various observers, however, under favourable conditions of the atmosphere and geographical position, have been able to trace the Band all the way across the heavens, thus showing that it apparently forms a complete arch from east to west. The existence of the Zodiacal Band is now generally accepted as amongst the known phenomena of the heavens.

The Gegenschein was first observed by the German astronomer Brorsen, in 1854. He appears to have given it the name it bears, which means "the reflection," or, literally, "opposite to the light." Hence the Gegenschein is frequently called the "counter-glow." The name, of course, bears reference to the light appearing in the part of the sky *opposite* the Sun. The Gegenschein has been independently discovered by different observers, unaware of it having been previously noticed. It was observed by Backhouse, at Sunderland, in 1871, and by Professor Barnard, at the Vanderbilt Observatory in North America, in 1883. These independent discoveries conclusively prove the existence of the counter-glow. Barnard further observed that it is larger and more diffused in the autumn than in the spring.

As regards visibility, the Gegenschein is so faint and indistinct in outline that it can be seen only under the most favourable conditions, including absence of moonlight, and its occurrence at a considerable altitude.. It is invisible when projected on the Milky Way. As we have noticed, it has been seen at Sunderland, but this is exceptional; and, on the whole, it can scarcely be supposed to be distinguishable under the usual atmospheric conditions obtaining in Great Britain. It is, however, somewhat brighter than the Zodiacal Band, and, as we have mentioned, its discovery was made by a German astronomer apparently observing in Germany.

The Zodiacal Band is so faint and uncertain that it can only be perceived with great difficulty under the best conditions. The atmosphere must be clear, and the observer's position favourable. Even under these circumstances its reality was doubted for a long time after its discovery, and it was only through the accumulation of observations that its existence came to be generally accepted. It seems scarcely probable that it can at any time be observed in our atmosphere.

The Zodiacal Light, although faint and inconspicuous, and so diffused as to be invisible through a telescope, is much brighter than the Zodiacal Band or the Gegenschein. Mr. Maunder, of the Royal Observatory, Greenwich, when on an expedition to India, observed it as being three times as luminous as the Milky Way. It is visible in Britain when the sky and atmosphere are clear, and moonlight absent, after twilight, in the evenings of the months of March, April, and May, in the western sky; and, before twilight, in the mornings of September, October, and November, in the eastern sky. It will be noticed that the Earth, in the interval of six months, passes to the opposite part of its orbit, so that the light is seen in the same part of the heavens on both occasions, although, during the former period, it appears on the western horizon, and, in the latter period, on the eastern.

The reason why the Zodiacal Light is seldom, if ever, visible in Britain, except about the time of the vernal equinox shortly after sunset, and about the time of the autumnal equinox shortly before sunrise, is owing to no change in the Light itself, but merely to our changing position through the orbital movement of the Earth. The Light lies in the Zodiac, with its axis in, or near, the ecliptic. In these northern latitudes, the ecliptic, in its stretch upwards from the western horizon at the hour of sunset, has, at the spring equinox, a more vertical character than at other seasons. This arises from the fact that the part of the ecliptic nearest to the north celestial pole—the place of the Sun at midsummer—is, at the time of the vernal equinox, on the meridian at

six p.m., that is to say, at the hour of sunset. The part
of the ecliptic on the meridian is thus about 23½ degrees
north of the celestial equator, and has consequently a
very considerable elevation in our latitudes. As the
position of the Sun, as it sets at the vernal equinox, is
on the celestial equator, the ecliptic, as it extends from
the western horizon to the meridian, advances in the
direction of the north celestial pole no less than 23½
degrees in the stretch of 90 degrees lying between the
part of the horizon at which the Sun has set and the
meridian. The ecliptic, therefore, as seen in Britain in
the early spring, is, as it rises from the western horizon,
not very greatly removed from the perpendicular.
Another factor which conduces to the Zodiacal Light
being visible in our latitudes, after sunset, at and about
the time of the vernal equinox, is the more direct
apparent diurnal course of the Sun at that season.
The Sun is then on or near the celestial equator, and
therefore rises due east and sets due west, or very nearly
in these directions. It thus comes about that the Sun
then sinks most rapidly below our horizon. Its apparent
course as it sets is more nearly vertical than at other
seasons. Thus we have the concurrence, at the same
season, of a rapid rise of the ecliptic from the western
horizon, and a comparatively short duration of twilight.
It is these circumstances together which account for the
fact that, at and near the time of the spring equinox, the
Zodiacal Light extends in its brighter portion above the
mists which, in these parts, usually occupy the horizon
at the place of sunset, and continues visible for a considerable time after twilight has disappeared.

Exactly the same state of matters occurs at the

N

autumnal equinox shortly before sunrise. The most northerly part of the ecliptic is then on the meridian not at six p.m.—that is to say the hour of sunset—but at six a.m.—that is, the hour of sunrise. An analagous state of matters occurs in the corresponding southern latitudes, but, of course, the times are reversed; the hours for observation being the morning hours at the vernal equinox, and the evening hours at the autumnal equinox. In lower latitudes, where the angle between the ecliptic and the horizon is greater than with us, and the duration of twilight shorter, the Zodiacal Light may be seen all the year round, provided the sky is clear, and that there is no moon, nor any conspicuous brightness from the planet Venus.

Various theories have been suggested as explanatory of the Zodiacal Light, the Zodiacal Band, and the Gegenschein.

We have seen that Kepler thought the Zodiacal Light formed the Sun's atmosphere, and that Laplace proved this to be untenable. We have seen that the Rev. George Jones thought the Light was caused by a ring of nebulous matter around the Earth, lying between the limits of our atmosphere and the Moon; a supposition negatived by Proctor and others. Dr. Hooke, a distinguished secretary of the Royal Society, in the latter part of the 17th century, suggested that the extra heat received by the Earth when nearest the Sun was given off as a luminous tail, which extended behind the Earth in its orbital journey when farther away from the Sun, and that this tail appeared to us as the Zodiacal Light. Cassini thought the Light might be produced by the reflected light of innumerable little planets circulating around the Sun. He and others noticed a resemblance between the Light and the faint luminosity of the tails of comets, and it has been suggested that the ether may be loaded with the materials of the tails of numerous comets. Sir William Huggins thinks it not improbable that the Zodiacal Light may be in some way connected with the outflow of extremely minute particles from the Sun; while the late Sir William Siemens thought an explanation of the Light might be found in the dust which he supposed to be ejected from the equatorial regions of the Earth, the dust being rendered luminous partly by reflected sunlight, partly by phosphorescence, and partly by electrical action.

It has been doubted whether the Gegenschein is really related to the Zodiacal Light. The difficulty of satisfactorily accounting for the light and the Gegenschein by a common cause has resulted in various explanatory hypotheses being put forward in relation to the Gegenschein as a separate phenomenon. It has been suggested that the Earth has a permanent luminous tail composed of the molecules of gases cast off from the outermost bounds of our atmosphere, and that these are repelled from the Sun by solar repulsion in the same way as the tail of a comet is driven away from the nucleus. This view is negatived by the absence of any indication of a shadow of the Earth in the Gegenschein, and the absence of any noticeable parallax. The suggestion has, however, been thought of some weight by scientists, as the quantity of matter necessary to produce such a tail would be extremely minute, and any such appendages on the other planets of the solar system would be quite invisible to us. Dr.

Moulton, of Chicago University, has shown by mathematical calculations in regard to the relative influence on each other of three bodies, such as the Sun, the Earth, and a minute body outside the Earth's orbit, that the last might be expected, under certain conditions, to make one or more small curved circuits in its orbital course when the Earth came directly between it and the Sun. Were a large number of particles similarly oscillating at once, a reflected light like the Gegenschein might appear.

The recent tendency has clearly been to seek the explanation of the Zodiacal Light, the Zodiacal Band and the Gegenschein in a common cause; and, although these phenomena are still considered mysteries, scientific opinion has come more and more towards the idea that the three phenomena may be explained through the existence of a mass of small bodies circulating around the Sun, and reflecting varying amounts of light upon the Earth. Professor C. Michie Smith thinks that there can be very little doubt that the explanation must be looked for in this direction; and the Rev. Edmund Ledger indicates the same opinion.

Notwithstanding this, however, no satisfactory conclusion has yet been come to, and Professor Simon Newcomb, writing in 1902, expressed the view that "of the various explanations which have been propounded no one can be considered as sufficiently probable to merit acceptance."* The Rev. Edmund Ledger thinks that although the Zodiacal Band and the Gegenschein may be satisfactorily accounted for by the supposition of the existence of a mass of small bodies circulating in

* Ency. Brit. XXV. 736.

the solar system, a difficulty occurs in relation to the Zodiacal Light itself. Why should it exhibit any definitely located apex or top, at the times when it extends to more than 90° from the Sun, and therefore beyond the Earth's orbit, as is often distinctly noticed to be the case ?

There can be no doubt that of all the theories propounded as explanatory of the various phenomena with which we are dealing, that which suggests the explanation to be reflected sunlight—the reflection being occasioned by a mass of small bodies circulating around the Sun and extending from comparatively near the Sun to outside the orbit of the Earth—has the greatest elements of probability and appears most satisfactory to reason.

Let us then examine this theory more carefully and see whether it is in any way supported by extraneous facts, and whether it satisfactorily harmonizes with the various luminosities which appear in the Zodiac.

Is there any reason to suppose that there exists a mass of small bodies moving in the solar system—partly at least in the neighbourhood of the Earth's orbit—apart from such a supposition being explanatory of the Zodiacal Light and relative phenomena ?

We have positive proof that such is the case.

The late Professor H. A. Nicholson, of Aberdeen, has given us, in an article in the *Encyclopædia Britannica*, much interesting information on the impact of meteors and meteorites on the Earth. He says :—

"Of the larger meteors there are in the mean six or eight per annum which in the last fifty years have furnished stones for our collections. A much larger number have doubtless sent down stones which have never been found.   Thus, Dubrée estimates for the whole Earth an annual number of six or seven hundred stone-falls.   But, of the small meteors or shooting stars, the number is very much larger.   Any person who should, in a clear moonless night, watch carefully a portion of the heavens, would, in the mean, see at least as many as eight or ten shooting stars per hour.   A clear sighted and practised observer will detect somewhat more than this number.   Dr. Schmidt, of Athens, from observations made during seventeen years, obtained fourteen as the mean hourly number on a clear moonless night, for one observer, during the hour from midnight to one a.m.   A large group of observers, as has been shown by trial, would see at least six times as many as a single person.   By a proper consideration of the distribution of meteor paths over the sky, and in actual altitude in miles, so as to allow for mists near the horizon, it appears that the number over the whole globe is a little more than ten thousand times as many as can be seen in any one place.   This implies that there come into the air not less than twenty millions of bodies daily, each of which, under very favourable conditions of absence of sunlight, moonlight, clouds, and mists, would furnish a shooting star visible to the naked eye.   Shooting stars invisible to the naked eye are often seen in the telescope.   The number of meteors, if these are included, would be increased at least twenty-fold."

It has been estimated that the actual weight of the meteors swept up by the Earth in each day's orbital journey is from 100 to 200 tons.   Mr W. F. Denning, the President of the Liverpool Astronomical Society, thinks it probable that more than 100 meteor showers, of very feeble character, occur every night of the year; and this does not take into account the meteors which fall to the

Earth during the morning and day time, and which, consequently, are unobserved.

Some of the meteorites which fall on the Earth are of very considerable size. Pliny mentions one which fell at Ægostopani, in 467 B.C., as being "as large as a waggon." In 1492, one fell in Alsace which weighed 270 lbs.; in 1627, one fell in Provence, which weighed 59 lbs.; and in 1873, one fell in California which weighed 12 lb., and penetrated the Earth to a depth of eight feet. Most meteorites, however, reach the Earth only as comparatively small stones or dust. It is evident from the great number of meteors which constantly fall to the Earth that, notwithstanding its great age, our planet is still in a state of "growth." Professor Nicholson considered the question of the density of meteors (or—as, for the sake of distinction, he calls those small bodies which do not enter our atmosphere—"meteoroids"), in the surrounding space. He concludes that the average number in the space the Earth traverses is, in each volume equal to that of the Earth, about 30,000. He considers that, on an equal distribution, each meteor, large enough to make a shooting star visible to the naked eye, would be about 250 miles from its neighbours. This is the estimated density applicable to the course of the Earth's orbital journey. The density of meteors in other parts of space is a mere matter of inference therefrom. There can, however, be no possible doubt that the solar system, both within and without the Earth's orbit, is occupied, to a great extent, by an innumerable mass of small bodies, each moving in its own orbit around the great centre of the system.

This, then, being so, let us consider whether the

appearance presented to our view by those meteoric bodies will satisfactorily account for the luminosity of the natural phenomena with which we are dealing.

We shall first turn our attention to the appearance of the western horizon immediately after sunset.

As the Sun disappears below the horizon, a great density of those small bodies will lie in the line of sight above the horizon, including not only those lying within the orbit of the Earth but those beyond our orbit. We may suppose that, as we gaze towards the western horizon, we have, in the line of sight, (1) those bodies situated between the portion of the terrestial orbit which the Earth is at the time passing through and a line extending upwards from the Sun; (2) those bodies situated between a line passing upwards from the Sun and the part of the terrestial orbit which the Earth will reach about eight or nine months later; and (3) those bodies beyond the part of our orbit which we shall reach in about eight or nine months, and between it and the boundary of the ring of bodies—or between it and the extreme distance at which the reflected light of such minute particles can be visible from the Earth.

The greatest density of the bodies, or particles, will thus, as viewed from the Earth, be nearest to the Sun; and the greatest luminosity will appear as near to the Sun as is consistent with the faint light not being lost in the glare of the Sun. In this line of sight the bodies situated nearest to the Earth will have their reflecting side almost entirely turned away from the Earth. They are in the position of the Moon when from one day to three or four days old, while those bodies at the greater distances will have, practically, their full luminosity

turned towards the Earth. The degree of luminosity of
the particles will thus vary according to position, from
the very partial illumination of the bodies near us to
the complete illumination of those at a great distance.
This, of course, refers to the *depth* of the ring of
particles near the Sun in the line of sight. To our view
all the particles in this line would appear—could we
distinguish them to any extent separately—to be the
same distance from the Sun. Although their distance
from the Sun will vary enormously, they set—or
disappear below our horizon—at the same interval after
the Sun. They have what is called the same "elongation," eastward of the Sun.

Let us now turn our attention to the *breadth* of this
ring of small bodies—that is, its extent northward and
southward. The part we have dealt with, in considering
the *density* of the ring, will, in relation to the *breadth*,
be the position of the central line.

The breadth, or width, of the Zodiacal Light, at its
visible base, which may be at an elongation of perhaps 20
degrees from the Sun, has been variously observed as
being from 20 to 30 degrees, or even more. Even the
smallest of these observed widths is somewhat greater
than the width of the belt of the sky which is known
as the Zodiac. The Zodiac extends about eight or nine
degrees on each side of the ecliptic, or the Earth's path
around the Sun, the ecliptic being the central band of
the Zodiac. The name Zodiac was given to this belt
from the figures of the constellations which lie in it
being for the most part called after animals. In this
belt there occur the movements of all the great planets
and of the Moon, as well as the apparent movements of

the Sun. Several of the minor planets, however, which
have been discovered in modern times, move in orbits
which extend outside this comparatively narrow belt;
the orbit of the asteroid *Pallas*, for instance, having an
inclination of no less than 35 degrees to the plane of
the ecliptic. This, however, is an extreme case, and, in
general, the paths of the asteroids do not deviate far
from the place of the ecliptic, the average deviation
being, it is supposed, only about 6 degrees.

Still, if the term Zodiac is used as descriptive of the
belt of the sky in which the movements of the planets
as now known (including the asteroids) occur, it must
be considered as having a width of about 35 degrees
on each side of the ecliptic. Thus, if the ring of small
bodies with which we are dealing has the full width of
the "Zodiac," using the term in this sense, it would,
if visible in its entirety, present to our view on the
western horizon, shortly after sunset, a width of about
70 degrees.

Although this is so, we are probably justified in concluding that at a distance of, say, 10 or 20 degrees from
the plane of the ecliptic, the number of the particles revolving in orbits around the Sun would be greatly reduced
from the number so moving with a less inclination, in
the plane of the ecliptic. All that is known as to the
revolution of the planets in the solar system supports
such a conclusion. The width of a ring of such particles,
in so far as in any excessive density, would, therefore, probably be not more than from 10 to 20 degrees on each side
of the ecliptic. This would give a total width of from 20
to 40 degrees. We must now notice that the position of
the particles in what may be called, the *line of width*,

just as in the line of depth, will affect their reflection of light upon the Earth, although to a smaller extent. The particles on what we may call the "borders" of the ring will be, to our view, somewhat north or south of the position of the Sun, and it is evident that the light reflected by these particles upon the Earth will somewhat diminish with their distance from the middle line, that is to say, will vary inversely with their distance north or south of the ecliptic, assuming the ecliptic to occupy the central line of the Light as it extends upwards from the horizon. In consequence of this, and of the lessening number of the particles with the increase of distance from the ecliptic, the light reflected would, at a certain (perhaps variable) distance north and south of the ecliptic, be quite imperceptible.

Let us endeavour to realize how the particles would appear to our sight to be affected by their *apparent* distance from the Sun. The apparent distance is, of course, to be distinguished from the real distance. We have seen that particles may appear to us to be near the Sun, while, in reality, they are far beyond the Sun, and beyond the opposite part of the Earth's orbit. The *apparent* distance is the elongation east of the setting Sun or west of the rising Sun—the interval between the setting or rising of the particles and of the Sun.

If we consider these minute bodies as forming a ring, or ellipse, extending out beyond the orbit of the Earth, we find that, as the apparent distance from the Sun increases, the density of the particles presented to our view must also decrease. Shortly after sunset—say at the equinox, being about six p.m.—there lies before us, on the western horizon, a depth of particles extending right across the space enclosed by the terrestrial orbit to, beyond the part of our orbit which we shall occupy about seven or eight months later. At midnight, on the other hand, our prospect on the western horizon is limited to the particles lying *outside* our orbit. Those within our orbit are altogether concealed below the horizon.

Let us imagine a broad, flat, iron ring, somewhat like the rings used in the game of quoits, and suppose the inner circumference to represent the Earth's orbit, and the enclosed space (excepting only a portion at the centre), as well as the whole area of the ring, to be occupied by these small bodies, more or less densely distributed. The Sun we shall imagine to be about the centre of the ring. It is clear that if the Sun were non-existent the greatest depth of the particles, as viewed from the Earth in any part of its orbit, would be straight across the centre of the ring. The depth of the particles would then be practically the diameter of the ring *plus* the width of the opposite part of the circumference. With the Sun in the position specified, and thus hiding the particles beyond it, the greatest depth will be the positions near the Sun. Such are the positions on the horizon after sunset and before sunrise. At midnight the horizon is represented by an arc of the circumference. The whole inner space is shut off, and we can merely look along the ring so far as this can be done in an absolutely straight line.

It is clear that, in the aspect of the horizon at midnight, not only is the density of the particles greatly lessened, but the amount of light received and reflected by each particle is also lessened. The particles at this

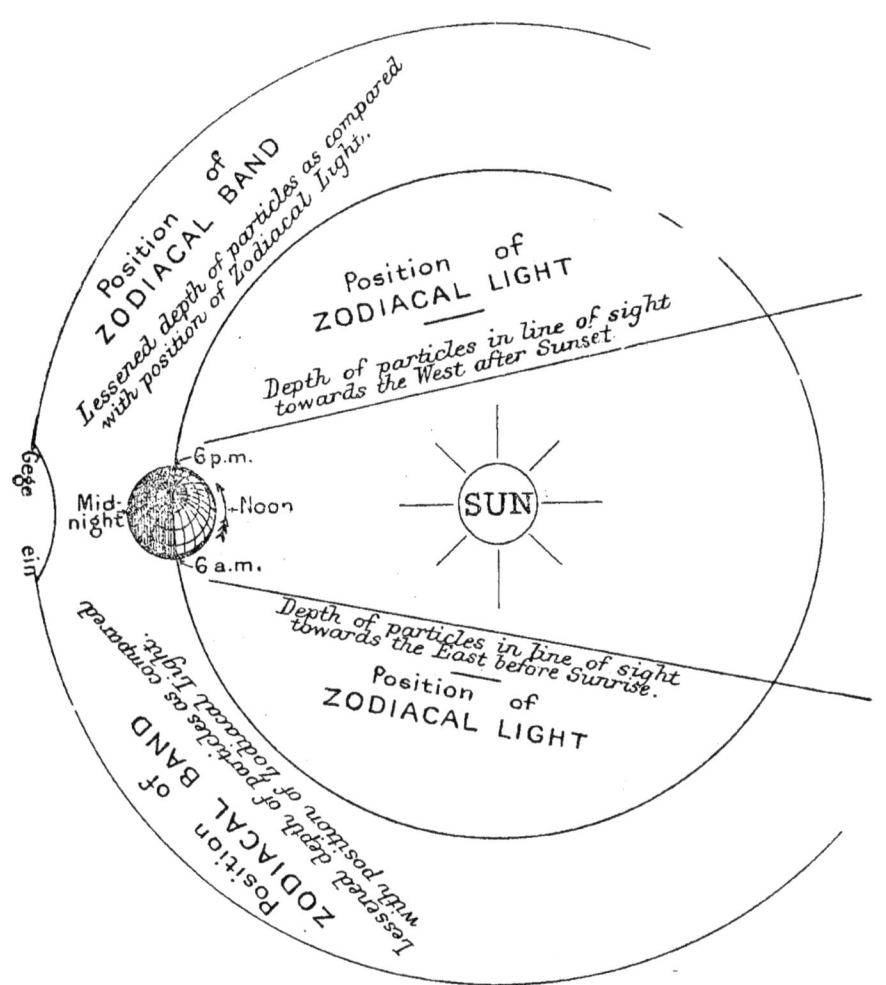

Diagram showing the varying density of meteoric bodies according to the direction of the line of sight and the hour of the day.—p. 190.

distance—being 90 degrees of elongation from the Sun—
are in the position of the Moon at first quarter, when one-
half of the illuminated disk is turned earthwards.   Near
the Sun there will be not only the greater density of
particles, but they will also, to a great extent, be fully
illuminated on the side turned towards the Earth.   The
particles on the horizon at mid-night are not only at a
great distance from the Sun—being all either outside, or
perhaps, to a slight extent, actually *on* the Earth's orbit,
but, to our view, they are all only partially illuminated.
In some degree also the light received by them may be
impeded by the intervening particles.   In fact, the
number of particles intervening must in this position be
actually the maximum.

If we continue beyond 90 degrees of elongation from
the Sun, we find that the density of the mass of particles
in the line of sight continues to decrease, but the portion
of the illuminated side of each particle turned towards
the Earth now increases.   If we use our illustration of an
iron ring, the least density of the particles in the line of
sight will be when we look from the inner edge of the
ring towards the outer edge.   The density is then merely
the breath of the circumference.   It will be noticed, however, that, in this position, the particles will have their
illuminated side turned fully towards the Earth.   They
are in the position of the full Moon.   Not only so, but
the fully illuminated particles are actually nearer to the
Earth than in any other position.   This is the position of
the Gegenschein.

It has been suggested by certain authorities that, if the
Zodiacal Light results from the reflected light of minute
bodies in the manner indicated, it would not exhibit a

definitely located apex when it is seen extended more than 90 degrees from the Sun. In this position the apex must be outside the Earth's orbit. A definitely located apex has, it is stated, been noticed at a greater distance than 90 degrees from the Sun, forming a contrast to the faint extension of the Zodiacal Band. It would seem that such an apex might reasonably be looked for, in view of the evident change of aspect of the ring of small bodies, *at a distance of about 90 degrees* from the Sun, and it would appear to be by no means impossible that it might be seen at a somewhat greater distance. Reverting to our illustration of a flat iron ring, it will be noticed that, beyond 90 degrees distance from the Sun, the particles in view are only those on the ring itself, and that, even as regards these, their density in the line of sight, and, therefore, their luminosity, lessens rapidly beyond 90 degrees until the margin of the Gegenschein is reached when other factors come into play. While the position of the apex should therefore naturally be at about 90 degrees from the Sun, its visible appearance will depend on atmospheric and other causes, and may be at a less or, quite possibly, at a somewhat greater distance.

All the large planets have a certain *eccentricity* of orbit, that is to say the Sun does not occupy the centre of the orbit. In the case of the Earth, the Sun is more than one-and-a-half millions of miles distant from the centre of the orbit, and from this it results that, when the Earth is in perihelion, or nearest to the Sun, the Sun is more than three millions of miles nearer to the Earth than when the Earth is in aphelion, or farthest away from the Sun. The eccentricity of the orbits of the asteroids, or small planets revolving between the orbit of *Mars* and that of *Jupiter*, varies greatly, varying from ·04 to ·34 of mean distance of the asteroid from the Sun. As the eccentricity of the Earth's orbit, although it results in a difference of over three millions of miles between perihelion distance and aphelion distance, is only ·0168 of our mean distance, it will be seen that, when the eccentricity amounts to ·34 of a mean distance greater than that of the Earth, the variation of the distance from the Sun will be excessive. With such an eccentricity, the variation in the Earth's distance from the Sun would be more than sixty millions of miles. It is supposed, and it seems probable from the analogy of the asteroids, that the eccentricity of very many of the small bodies with which we are now dealing will be very considerable,—that their distance from the Sun in different parts of their orbit will vary to a great degree. From this it may result that, in certain parts of its orbit, the Earth may have comparatively few particles, or perhaps none, opposite the place of the Sun. The Gegenschein may actually not exist, the Zodiacal Band may be broken.

Professor Newton, of Yale University, who has given much study to the subject of meteors, satisfied himself that, in the great majority of cases, the meteorites which had been actually seen to fall upon the Earth must have moved in orbits inclined less than 90 degrees to the plane of the Earth's orbit, and that their perihelion distance lay between that of the Earth and one-half that of the Earth. Thus, these meteorites would, at no time, be nearer the Sun than say 44,948,500 miles, that being about half the Earth's perihelion distance. This would place them outside the orbit of *Mercury*, whose greatest distance from the Sun is about 42,669,000 miles. This, however, affords no reason for supposing that these small bodies do not occur inside the orbit of *Mercury*. So far as the argument goes, it suggests merely that those members of the meteoric system which cross our orbit do not approach nearer to the Sun than one-half of the Earth's nearest distance. It is not suggested that there are not vast numbers of the particles much nearer to the Sun ; but that these in their orbital journey are mostly, even when at their greatest distance from the Sun,—that is in aphelion—still nearer the centre of our system than the nearest part of the Earth's orbit.

If, as we have seen there is reason to suppose, many of the meteoric bodies have very eccentric orbits, so that in one part of their orbital journey they are outside the orbit of the Earth, and at another part within our orbit, there must, at parts of our orbit, be many of these particles crossing from outside our orbit to inside our orbit. Of course, they might so cross—as many of them must doubtless do—on a different level from the Earth. They may be what we may describe as "higher" or "lower" than the Earth. If, however, the particles are to any extent situated on the plane of the ecliptic at crossings of our orbit, then the risk of collisions between the particles and the Earth will be considerable. If the particles are, as is supposed, virtually innumerable, and if they are so distributed that there is no considerable portion of the space in the neighbourhood of our orbital course free from them, then there must always—if their orbits have such eccentricity as we have described—be some of the particles at these crossing places. This being so, we are certain when passing over such parts of our orbit to collide with the particles. We may suppose that certain parts of our orbit are common crossing places of the more noticeable particles. In these parts large numbers of such particles will be encountered. We may suppose that the portion of the orbit which we pass over in November is such a part, and that what we have come to call the November Meteors are simply some of the larger particles then crossing our path in their own orbital journey. We may also conjecture that, at certain parts of the ring, the larger particles are more crowded together than at other parts, that there is a greater density in a given area. The crossing of our orbital path by such a crowd of particles would explain the great showers of meteors which occur at somewhat irregular intervals.

If the Gegenschein is produced by reflected sunlight, from small bodies situated outside the orbit of the Earth, in a position opposite to the Sun, it must follow that part of the light of the Sun will be intercepted by the Earth, so that in the centre of the Gegenschein it would seem that the shadow of the Earth should appear. There seems, however, to be no trace of the shadow.

The explanation probably lies in the fact that the particles are sufficient in density to reach beyond the shadow cone. The extreme length of the Earth's shadow cone, which occurs when the Earth is at its greatest distance from the Sun, is 870,300 miles, and the shortest length, which occurs when the Earth is in perihelion, is 843,300 miles, the average length of the Earth's shadow being 856,800 miles. If, therefore, the outer boundary of the ring of small bodies is at a greater distance outside the Earth's orbit than 870,300 miles, as an extreme, the Earth's shadow would not reach it,—the light from the Sun would not be intercepted. It might be supposed that, even in these circumstances, the central part of the Gegenschein should show diminished luminosity, from the fact that a lessened density of particles would be reflecting light, the nearer bodies having the light cut off from them through the intervention of the Earth. It is apparent, however, that those particles in the central portion of the Gegenschein are, in so far as the light of the Sun is not intercepted, precisely in the position to give the most brilliant reflection. Although the particles within the distance of the shadow cone will have their light intercepted, those beyond that distance will be in the full glare of the Sun, and be fully reflecting that glare upon the Earth. These particles which are in the line of extension of the darkest shadow are in the position of fullest luminosity. If, however, at certain parts of our orbit the outer boundaries of the ring are narrowed, and come within the distance to which the Earth's shadow extends, it might be expected that the shadow should be distinguishable in the centre of the Gegenschein. It would seem that, even in these circumstances, the shadow will be virtually undetectable. The light of the Gegenschein is itself so faint and uncertain, that it is only seen with difficulty under very favourable conditions. Even practised observers have doubted their own observations in regard to it. This glimmering luminosity is of great size, but, even in regard to its size, the results of observation are very indefinite; the diameter, as observed, varying from 10 degrees to 20 degrees. Now, the shadow of the Earth, even at the mean distance of the Moon, (238,818 miles) would have a diameter of probably not more than about one degree, and, at a greater distance, the diameter of the shadow continuously decreases till it reaches a fine point.   Now the darkness, and, therefore, the visibility, of a shadow bears a definite relation to the object against which it is cast.   A shadow appearing in contrast to a glaring light is very conspicuous, but, in contrast with a light which is itself almost indistinguishable, the shadow is dissipated.   It can scarcely be supposed that, against what we may call the shadowy glimmer of the Gegenschein, the slight shadow of the Earth can, even under the best conditions be observed.   At least, we should imagine that prolonged and delicate observation may be necessary before this result can be attained.

Let us turn our attention to the converse of the Gegenschein, and see whether the existence of these minute bodies between us and the Sun can be reconciled with the fact that there is no apparent interception of the light of the Sun.   How can it be the case that the space between the Earth and the Sun is occupied by a mass of minute bodies without the Sun itself being obscured and the light of the Sun intercepted ?

The *existence* of these small bodies in the space outside the Earth is conclusively proved; and it can scarcely be doubted that the space between the Earth and the Sun, in our line of sight, is densely occupied by them.   That they must, to a certain extent, intercept the full glare of the Sun seems highly probable.   It may, however, be conjectured that the extent of the obscuration will not be more noticeable than the light of the Gegenschein itself.   As the light of the Gegenschein is exceedingly faint, and is with difficulty observed, so the amount of sun-light cut off from the Earth by the "celestial dust" of inter-solar space will, in comparison with the glare of the Sun, be quite inappreciable.

From the fact that the asteroids (which are, certainly, very much larger than the small bodies with which we are dealing), are only being gradually discovered by patient investigation at well equipped observatories, it would appear that the particles floating in space between the Earth and the Sun must be quite invisible even through the largest telescopes, so that their transit across the Sun's disk would not be distinguishable.

Some observers have asserted that they have been able to distinguish a light in the neighbourhood of the Moon similar to the Zodiacal Light. They have suggested that the Moon produces an appearance very like the appearance of the Zodiacal Light. There appears to be some doubt as to whether this suggestion can be accepted. The matter is referred to by Professor Michie Smith in an article on the Zodiacal Light. He mentions that one observer, Mr L. Trouvelot, made an interesting observation in this connection.

"On a night when the Zodiacal Light was very bright, and there were magnetic disturbances followed by an auroral display, but when no aurora was actually visible, he saw a conical light rising obliquely from the top of the roof of a building behind which the Moon, then about 15° or 20° above the horizon, was concealed. The axis of the light coincided nearly with the ecliptic, and the light could be traced on both sides of the Moon when the Moon itself was concealed. The whole of the circumstances led Mr. Trouvelot to conclude that this light and the Zodiacal Light were phenomena of the same order, while this and other observations, he considered, rendered it probable that there was some connection between the Zodiacal Light and auroras."

Although this appears to be a very exceptional observation, and the existence of a light similar to the Zodiacal Light in relation to the Moon is not yet accepted, there would seem to be no inherent improbability in the supposition. If the Zodiacal Light is the reflection of sunlight by small bodies moving in the Zodiac, it is quite conceivable that, under favourable circumstances, the light cast by the Moon—although itself merely reflected light—upon the small bodies in its neighbourhood might be visible to careful observers. It will be noticed that this light was observed only while the Moon was concealed, so that it would appear that the light of the Moon when in view hid the exceedingly faint light observed by Mr. Trouvelot. His observation is quite consistent with the theory of the Zodiacal Light being the Light reflected by numerous small bodies exposed to the brightness of the Sun. It does not appear what was the age of the Moon when this observation took place. If the Moon were full, it would then have been situated in the Gegenschein. If it were not full to our view, it would, nevertheless, have appeared as fully illuminated from the position of certain of the particles. In this case, the light reflected by the Moon would be thrown chiefly towards the illuminated side. The information is, however, very indefinite.

We have now seen that the Zodiacal Light, the Zodiacal Band, and the Gegenschein, appear to be capable of explanation by the supposition of the existence of a ring or ellipse of small bodies extending outwards from near the Sun to beyond the orbit of the Earth, its plane approximating to the plane of the Earth's orbit, and that there is strong reason from extraneous evidence to believe in the *existence* of such a ring or ellipse.   We have seen that the Earth in its orbital passage constantly encounters small bodies in the form of meteorites, sometimes as isolated fragments, sometimes in showers; that not a day passes without thousands—perhaps millions—of minute particles being encountered.   We have seen that these particles are, in all probability, the cause of the Zodiacal Light and the allied phenomena, so that, in the numerous specimens of meteoric stones in our museums, it may be that we are actually able to handle and examine the substances which, even in our own day, may have occasioned a fragmentary part of the Light which appears in the Zodiac, under the varied form of the Zodiacal Light, the Zodiacal Band, and the Gegenschein.

THE WINDS—THEIR CAUSES AND CHARACTERISTICS.

SYNOPSIS.

The winds as represented in mythology—Definition of the wind—Principal causes of the winds—The trade winds—The monsoons—Prevailing winds in tropical and temperate zones — Noticeable local winds—Variation in force of wind at different levels—Sea and land breezes—Prevailing winds of the British Isles—Cyclonic and anti-cyclonic movements of the atmosphere—The law of the winds—The rotatory character of the winds in cyclones and anti-cyclones — Diverse movements in northern and southern hemispheres—The poles of the prevailing winds—Causes of dryness or humidity of winds—Rainfall in relation to wind—Velocity and relative pressure of winds—Wind pressures and velocities in storms—Cyclones in tropics and higher latitudes compared—The deviation of the wind from the horizontal—Cyclonic and anti-cyclonic movements considered in relation to the horizontal angle of the winds—Effect of local physical characteristics on the angle of the winds—Importance of observation of the winds' deviation from the horizontal and the present lack of scientific apparatus for the purpose.

THE WINDS—THEIR CAUSES AND CHARACTERISTICS.

IN mythology each of the winds is endowed with a personal existence and human parts and passions. Æolus, the god or king of the winds, is represented by Virgil, in the Æneid, as enthroned in a cavern controlling his troublesome subjects. The passage has been well translated by Conington, who was Professor of Latin at Oxford, for about fifteen years, up to his death, in 1869.

> "Here Æolus, in cavern vast,
> With bolt and barrier fetters fast
> Rebellious storm and howling blast.
> They, with the rock's reverberant roar
> Chafe, blustering round their prison door ;
> He, throned on high, the sceptre sways,
> Controls their mood, their wrath allays.
> Break but that sceptre, sea and land
> And heaven's etherial deep,
> Before them they would whirl like sand,
> And through the void air sweep."

The character of each of the winds is typified by its mythological representative. Thus the blustering north wind is represented by rude Boreas with distended cheeks blowing his spiral shell; the pleasant west wind by gentle Zephyrus with lap full of flowers; the rainy south wind by Notus bearing a water jar.

These poetic ideas have no place in our prosaic times in the study of meteorology. It has, indeed, been suggested that the conception of the god of the winds may have a basis of fact. Certain writers consider that Æolus may have had a real existence, and that the power over the winds, which is credited to him, had its origin in his skill in meteorology, and the management of ships. It is thus quite conceivable that this mythological fancy is based upon the history of a master of meteorology of a time long antecedent to the period of the author of the Æneid. Possibly, in our investigations into weather phenomena, we are only groping our way towards discoveries which were made in long past ages, and were subsequently lost.

The wind may be defined as a movement of the atmosphere which does not correspond with the movement of the neighbouring land surface or water surface of the Earth. If the movement of the atmosphere, as a whole, corresponded with the orbital and rotational movements of the Earth, there would be no wind. At the equator, the rotational movement of the Earth causes a surface displacement of about $17\frac{1}{3}$ miles a minute, so that the atmosphere at the equator will have a rotatory movement at the Earth's surface of $17\frac{1}{3}$ miles every minute without any wind being perceptible. If, however, the rotatory movement of any portion of the atmosphere on the Earth's surface in that region falls below, or rises above, the velocity mentioned, a wind will be occasioned. The Earth's rotational movement is towards the east, and, therefore, if any part of the atmosphere on the surface at the equator is rotating with less velocity than the Earth, a wind will there appear to blow from the east. If, on the other hand, the rotation of any part of the atmosphere is more rapid than the Earth, a wind will there appear to blow from the west. Persons in a balloon, being carried along by the atmospheric currents, do not, generally speaking, feel any wind, whatever the strength of the wind may be. We recognize the movement of the atmosphere only when it does not correspond with our own movements, whether the latter are occasioned simply by the movements of the Earth or by change of position on its surface. We have the converse of the balloon borne along by the wind, in the rapid passage of a railway train, or a motor car, in which, even in a calm, the occupants find an apparent breeze blowing in the opposite direction to that in which they are travelling. Thus, the wind is simply a movement of the atmosphere, which does not correspond with the movements of the neighbouring solid or liquid surface of the Earth; and an apparent wind is, similarly, an apparent movement of the atmosphere occasioned by an actual movement of the observers in the opposite direction.

The principal causes of the currents, or movements of the air, which we recognize as winds, are the unequal distribution of heat in the atmosphere and the axial rotation of the Earth. The air is a gaseous fluid, and its natural state is one of rest, and this state it always endeavours to preserve, and, if lost, to recover. The air at the surface when heated by terrestrial radiation becomes lighter and consequently rises to a higher level, while the colder air replaces it. The equilibrium of the air being thus destroyed, an atmospheric current is caused by the effort to restore it. The general fact which determines the occurrence of winds is indeed the law of gravitation. In accordance with this law, the heavier air seeks to pass under the lighter air, and the lighter air seeks to ascend till it finds its equilibrium. As the lighter air ascends it, in turn, loses its heat; as the heavier air descends it, in turn, gains heat from the Earth, and so the ceaseless movement of the atmosphere goes on.

The heating of the air is ultimately due to the heat of the Sun, and as this operates unequally, not only according to geographical position, but according to the season, and according to the time of day or night, a highly complicated effect is produced as regards the movements of the atmosphere.

The atmospheric currents which are recognized as being most directly connected with the axial rotation of the Earth are those which are known as the Trade Winds. These winds are constant on all open seas on both sides of the equator, extending to the distance of perhaps nearly 30 degrees north and south of it. The name has reference to the fact that their regular occurrence is taken advantage of in the interests of commerce. The direction of the Trade Winds differs on each side of the equator. North of the equator, they blow from the north-east, varying at times a point or two of the compass either way. South of the equator they proceed from the south-east, with a corresponding variation.

These winds are caused by the excessive heating of the air at the equator, and the Earth's rotation. Through the heating of the air—by terrestrial convection of the solar heat received by the surface—a column of air is, in these regions, caused to ascend from the Earth's surface. To replace this air, a current sets in towards the equator from the north and from the south. These currents might naturally be expected to be north and south winds respectively. This would, no doubt, be the case, but for the axial rotation of the earth. The displacement of any portion of the Earth's surface, caused by the axial rotation is greater in a given time at the equator than at any other part of the Earth. This arises from the length of the equator being greater than the length of any circle of latitude. The circumference of the Earth at the equator is about 25,000 miles, so that every portion of the Earth's surface at the equator is caused to pass through 25,000 miles daily, merely through the Earth's rotation. The length of the circle of latitude 30 degrees north, or 30 degrees south, of the equator is about 21,500 miles, so that in these latitudes every portion of the Earth's surface passes, by the diurnal rotation, through about 3,500 miles less than it would if situated on the equator. Between these latitudes and the equator, the distance through which every spot is carried by means of the diurnal rotation increases, while between these latitudes and the Poles it decreases, until, at the Poles themselves it is non-existent. Thus, a current of air passing from latitude 30°—either north or south— towards the equator, passes in the direction in which the axial velocity increases. The air so passing has acquired the axial velocity of the latitude which it is leaving, and, as it advances in the direction of the equator, it is left behind the rotating surface, and will thus seem to be moving in a direction contrary to the earth's rotation. As the Earth's rotation is from west to east, the current of air will seem to be from east to west. The air therefore, which is passing from the north towards the equator will have, in the first place, a movement from north to south, and, in the second place, a movement from east to west, and the air which is passing from the south towards the equator will have, in the first place, a movement from south to north, and in the second place, a movement from east to west. These movements combined give us, to the north of the equator, a north-east wind, and to the south of the equator, a south-east wind.

The varying position of the Sun has, of course, an effect on the Trade Winds. The Sun is directly above the equator only at the equinoxes. Between the vernal equinox and the June solstice, it apparently passes about 23½ degrees to the north of the equator. As the Sun passes farther and farther to the north of the equator, the south-east wind becomes gradually more southerly and stronger, and the north-east wind more easterly and weaker. The effect is reversed when the Sun in its apparent change of position is to the south of the equator.

The Trade Winds are constant only over the open ocean, and the larger the ocean the more steady they are. This is specially noticeable in the Pacific. When these winds blow over land, they are obstructed, and their direction is influenced by coming in contact with elevated lands.

The belt which lies between the two Trade Winds is characterized by calms, which, however, are frequently interrupted by violent storms.

The Monsoon is a Trade Wind subject to a seasonal

variation. The name Monsoon is derived from the Arabic word *mausim*, which means "a time" or "a season." The Monsoon, for six months of the year, blows in the same direction as the Trade Wind which prevails on the same side of the equator. For the remaining six months, however, it blows in the opposite direction. The Monsoon occurs in the neighbourhood of the equator, and is specially prevalent on or near land surfaces. It appears to have its origin in the changing relation of the Sun to the Earth's surface. Thus, when the Sun is considerably to the north of the equator, the surface beneath it may be heated to a greater extent than the surface at the equator, and thus the radiation will be greater in such a position than at the equator itself. As the heated air ascends, a current will thereby be occasioned from the direction of the equator northward, and this current will flow from a part of the Earth where the rotatory speed is excessive towards a part of the Earth where it is not so great. The atmospheric current will, consequently, have a movement from west to east greater than the corresponding movement of the surface in its neighbourhood. It will, in so far as it is affected by the rotation of the Earth, seem to blow from the west, and as its own independent movement is northward, it will form an atmospheric current blowing from the south-west. This will occur in a latitude in which the usual Trade Wind is from the north-east. An exactly analogous state of matters will, in the opposite season, be occasioned in positions on the south side of the equator. When the Sun is to the south of the equator, a current will be drawn from the equator which will blow as a north-west wind, the ordinary

Trade Wind of the latitude being from the south-east. As the Monsoon is chiefly prevalent on land surfaces, the direction of the wind is liable to be affected by the surface elevations, so that it may differ considerably from the north-easterly and south-westerly directions, which should naturally prevail. The radiation of heat being affected by the character of the land surface, the prevalence and strength of the Monsoon will depend, to a great extent, on the character of the land surface.

Dr. Alexander Buchan, the distinguished meteorological authority, states that the term "Monsoon" has long been applied to the prevailing winds in Southern Asia, which blow approximately from south-west from April to October, and from north-east from November to April; and that the term is now generally applied to those winds connected with continents which are of seasonal occurrence. As they are caused directly by the different temperatures and atmospheric pressures which form marked features of the climates of continents in winter and summer respectively, they are most fully developed round the coast of Asia, owing to the great extent of that continent. Dr Buchan continues :—

"The Monsoons of different parts of the coast of Asia differ widely in direction from each other. Thus in winter and summer respectively they are W.N.W. and E.N.E. at the mouth of the Amur; N. and S.S.E. at Shanghai; N.E. and S.W. at Rangoon; N. and W.S.W. at Bombay; N.W. and S.W. at Jerusalem; and S.S.W. and N.N.E. at Archangel. The Indian winter Monsoon generally begins to break up in March, but it is not till about the middle of May, when the normal pressure has been decidedly diminished over the heated interior, that the summer Monsoon acquires its full strength, and the heavy monsoonal rains fairly set in. In October, when the temperature has fallen considerably, and with the falling temperature the pressure of the interior has risen, the summer Monsoon begins to break up, and this season is marked by variable winds, calms, and destructive hurricanes. As the temperature continues to fall and pressure to rise, the winter Monsoon again resumes its sway. Monsoons, equally with the Trade Winds, play a most important part in the economy of the Globe. The relatively great force and steadiness in the direction in which they blow, and the periodical change in their direction, give facility of intercourse between different countries; and besides, by the rainfall they bring, they spread fertility over extensive regions which otherwise would be barren wastes."

From observations which have been made in numerous parts of the Earth, it has been ascertained that within the tropics the prevailing winds are easterly, while in the north and south temperate zones, they are westerly. In both cases the main cause of the easterly or westerly direction of the winds is the rotation of the Earth. As we have noticed, an aërial current proceeding from the higher latitudes towards the equator is advancing in the direction of greater rotatory velocity, and is thus left behind the Earth in its eastward movement, and is consequently felt as a current blowing from the east, while an aërial current proceeding towards higher latitudes is advancing in the direction of less rotatory velocity, and thus has a more rapid motion of rotation than the Earth in its neighbourhood, and is consequently felt as a current blowing from the west.

While the causes of the existence of the winds are as we have stated, their direction and character are considerably influenced by local features. In every situation particular geographical causes come into play in deciding the direction and nature of the prevailing winds, in addition to the great main causes of their origin.

Many of the more noticeable and violent local winds have received special names. Such are the Etesian Wind, the Mistral, the Simoon, the Typhoon, the Tornado, the Northers, the Pampero, the Zòbà'ah, the Gregàlia, and no doubt many others less well known.

The Etesian Wind—which has derived its name from the Greek word *etesios*, which means "annual"—is the name given by Greek and Roman writers to the periodical winds which occur in the Mediterranean.

The Mistral, which literally means the "master"—that is the master wind—is a violent, cold, north-west wind, which occurs in Provence, in the south-east of France, and in other districts bordering on the Mediterranean. It blows with greatest violence in the autumn, winter, and early spring, and is supposed to be due to the cold, dense air of the Alps and Cevennes rushing in to replace the heated air which ascends in these warm southern districts. The Mistral is one of the greatest scourges of the region. It destroys crops, fruits, blossoms, and vegetation, and is a terror to mariners in the adjoining sea.

The Simoon, or, as it is variously called, the "Simoom," or the "Samoom," is a name derived from the Arabic word *samma* "to poison." It occurs, and is known by one or other of the foregoing names, in Arabia and in Egypt, and neighbouring parts. The same or a similar wind is known by other names in various countries.

In southern Italy it is called *Sirocco* or *Scirocco*; in Turkey, *Sauriel*; in Spain, *Solano*; in Syria, *Khamsin*, or *Kamsin* (a name which also is given to it in Egypt); in Morocco, *Shume*, or *Asshume*, and in Guinea and Senegambia; *Harmattan*. It is a hot, dry, suffocating wind, which has its origin in the extreme heat of parched deserts or sandy plains. The air, heated by contact with the scorching sand, ascends in a great column, and the colder air from all sides rushes in, and forms a whirlwind, which is borne along laden with sand and dust. The intense dryness and heat of the wind, combined with the cloud of dust and sand which it carries along, render it most injurious to both vegetable and animal life. Its effects are felt in the neighbouring regions, where winds are occasioned through its influence, which are in turn considered as the winds of the desert. The effects are, of course, modified by the nature of the Earth's surface over which it passes.

The Typhoon is named from the Chinese words *tai-fong* or "great wind," the spelling being influenced by Greek *typhōn*—a "violent whirlwind." The Typhoon is a fierce hurricane of a class which occurs, not infrequently, in the months from May to November, on the coasts of China and Japan, and the neighbouring archipelago. These storms are most common and disastrous in July, August, and September.

The name Tornado is frequently applied to any tempest or hurricane, but it is specially given to the violent whirling wind or tempest prevalent about the time of the equinoxes, in the West Indies, and on the western coasts of Africa, and, about the changes of the Monsoons, in the Indian Ocean. Tornadoes are usually accompanied by severe thunder and lightning, and by torrents of rain,. but they are generally of short duration and restricted extent. In fact, what is called a Tornado in the West Indies, off Western Africa, and in the Indian Ocean, is called a Typhoon in the seas which skirt China and Japan. The name Tornado is from the Spanish word *tornada*, " a return," and has reference to the turning or whirling movement of the wind. Whirlwinds, waterspouts, dust storms, and tornadoes are essentially the same. They differ from each other only in their dimensions, their intensity, or the degree in which the moisture is condensed into visible vapour.

The Northers, or the Nortes, is the name given to certain violent gales from the north which prevail in the Gulf of Mexico, from September to March. They are peculiarly dry, cold, strong winds. When, as frequently happens, they follow a storm, their disagreeable qualities are intensified. The temperature falls rapidly from, perhaps, upwards of 80° F., to 18° F. or less, the wind in the meantime blowing fiercely. A violent wind with with such a low temperature as 18° F. is quite unknown in Britain.

The Pampero, or the wind of the " Pampas "—as the South American treeless plains are called—is a violent wind from the west or south-west, which sweeps over these plains. The Pamperos are supposed to be portions of the return, or north-western Trade Winds, and they are often felt far out at sea. They occur at all seasons, but most frequently from October to January. They are preceded by easterly winds ; and an uncertain fall of temperature with 'rising barometer and change of wind, are their constant and most striking characteristics. It is on record that on one occasion the temperature fell 44° F. in four-teen hours, while on another occasion, the fall was only four degrees. Rain is a usual, but not invariable, accompaniment of the Pampero.

The Zòbà'ah is a common but remarkable phenomenon in Egypt. It is a very lofty whirlwind of sand, re-sembling a pillar, which moves with great velocity. Sand columns generated by the wind were found, by measurement with a sextant, to be between 500 feet and 700 feet in height, and one was seen having an altitude of 750 feet. Boats are frequently capsized by the Zòbà'ah, as it sweeps across the Nile, especially if the main sheet is tied instead of merely being held by the hand.

The Gregalia, or the "Gregale," as it is called in Malta, is the wind through which St. Paul suffered shipwreck, and which he names "Euroclydon." It is a furious, dry, cold wind, which frequently blows in the Levant, or eastern part of the Mediterranean, for two or three days at a time, especially in November and February. It will be remembered that St. Paul was cast on the Island of Malta (or "Melita"), and it is in connection with that Island that this wind is chiefly known. Its direction is from the north-east or north-north-east.

It is believed that in the higher parts of the atmosphere the force of the wind is generally much greater than it is at the surface level. Balloons, at a height of two miles, have been carried along at a speed exceeding 60 miles an hour, while, at the surface, the force of the wind was not more than 15 miles an hour. This variation in the velocity of the wind at different levels, is also illustrated by the force of the wind on mountain tops, which is generally considerably greater than in the surrounding valleys. Of course, in the open sea, where the wind is unobstructed, its force is frequently much greater than on the broken land surface, the sea in this respect being comparable to a mountain top with an unobstructed exposure.

In the higher strata of the atmosphere the direction also of the wind may be quite different from that at the surface level. This is evidenced by the equatorial ascending column, which, on attaining its specific level, spreads northward and southward, while at the surface the inrushing air is in the direction of the equator.

The direction of the prevailing winds in any particular place is decided—subject to the general movements of the atmosphere, to which we have referred—by the geographical position of the place, and by the physical configuration and character of its surroundings. Thus, at all places near the sea coast we have what are called sea and land breezes, alternately by day and night; unless the tendency to the formation of these breezes is overpowered by stronger winds or obviated by other causes. During the day-time the temperature of the land is raised by the Sun to a greater extent than the temperature of the sea. The atmosphere resting on the land surface is thereby heated and rises, and the air from the sea rushes landward to replace it. In the night-time the state of matters is reversed, the temperature of the land surface falling more rapidly than that of the sea surface. The air above the land thus loses its heat and descends, and an atmospheric current is set up from the land to the sea.

The regular recurrence of sea and land breezes is dependent on the atmospheric pressure being tolerably equable, or the barometric gradient small, and the heat of the Sun moderately strong. Dr. Buchan mentions that these breezes occur in "even such an unsettled climate as that of Scotland when the weather conditions are favourable." He instances the coast of Berwickshire as affording, under suitable circumstances, a remarkable example of the veerings of the wind, owing to the alter-nating sea and land breezes.

"In the morning, the wind is north-west till about 10 a.m., when it veers to north, falling all the time till finally it sinks to a calm. A little before noon it springs up from north-east or east, veers to south-east from 2 to 3 p.m. where it continues till 7 p.m., about which time it veers to south, and then south-west, diminishing in force, and finally sinking to a calm. About sunset it springs up from west, veering to north-west during the night, where it continues till the following morn-ing. The wind thus virtually makes the round of the compass, is strongest from north-west and south-east, and weakest at north-east and south-west, being thus strongest when its course is perpendicular to the line of coast."

In Great Britain as a whole, the wind is more frequently from a south-westerly than from any other direction, but in April, May, and part of June, its direction is very commonly easterly, especially on the east coast. In Ireland, the south-westerly and westerly winds are the most prevalent.

The atmospheric current, which forms a surface wind, is generally, if not always, connected with either a cyclonic or an anti-cyclonic movement of the atmosphere.

A cyclonic area is a region in which the atmospheric pressure is low. There is, so to speak, a deficiency of air in the central portion, and consequently, a current sets in from all directions towards this central portion to make up the deficiency. An anti-cyclonic area is exactly the converse. It is a region in which the atmospheric pressure is high. There is a superfluity of air in the central portion, and a current flows outwards in all directions to get rid of the superfluity. In the centre of the cyclonic area there is an ascending column of air, so that the air is being withdrawn from the lower to the higher levels, and the air so withdrawn has to be made good out of the surrounding lower strata of the atmosphere. This is what occurs in the equatorial regions when the surface is greatly heated, and excessive radiation is set up. It is, indeed, what occurs on a minute scale on the coast daily, when the sea breeze is occasioned. In the centre of the anti-cyclonic area there is a descending column of air, and this air, as it reaches the surface and lower levels, flows outwards over the surrounding parts. On a minute scale this action is typified by the nightly occurrence of the land breeze on the coast. The worst summer weather in Great Britain is when low pressures—or cyclonic areas—prevail over the North Sea; and the hottest and most brilliant weather, when anti-cyclones lie over the land, and extend away southward and eastward.

The winds in anti-cyclonic areas, or regions of high atmospheric pressure, are usually very moderate, while in cyclonic areas, or regions of low atmospheric pressure, they are generally strong. Of course, this applies to the surface level. The most important difference, however, between cyclones and anti-cyclones consists in the direction of the wind, which, in the cyclonic area, circles inwards; in the anti-cyclonic area, outwards. Another important difference consists in the character of the air in the respective areas. Over a region covered by an anti-cyclone, particularly in the central portion, the air is very dry, and either clear or nearly free from clouds. The anti-cyclonic area is, on account of the dryness and clearness of the air, more fully under the influence of both solar and terrestrial radiation. Consequently, in winter, an anti-cyclonic region has great cold, with clear calm air, while in summer it has great heat, with probably a considerable fall of temperature during the night. The ascending column of air at the centre of the cyclonic area must necessarily spread outwards in the upper strata of the atmosphere, while the descending column of air at the centre of the anti-cyclonic area must necessarily, in the upper strata of the atmosphere, be fed by inflowing currents. The cyclone is thus in all respects the converse of the anti-cyclone, and these two diverse atmospheric movements may be regarded as the complement of each other.

Of course, a cyclone or an anti-cyclone, in general, covers a great extent of the Earth's surface, and it is only by observation over wide tracts of country that these atmospheric movements can be identified. Cyclones have diameters seldom less than 600, and occasionally more than 3,000, miles in length, and it is probable that anti-cyclones cover an equal area. Dr. Buchan, in reviewing the weather of north-western Europe for 1868, refers to an anti-cyclone which overspread the greater part of Europe, from the 2nd to the 4th of August in that year. He found that, in Iceland, it was bounded by part of a cyclonic system, and, in the Crimea, by another cyclonic system. It seems likely that a similar state of matters

is invariably the case, and thus that these great atmospheric movements are each in touch with movements
of a diverse character.

It is known that the wind does not blow directly from
a region of high pressure towards a region of low
pressure. We have seen that the actual local direction
of the wind is influenced by diverse causes, so that this
cannot occur. It has been found that, in the Northern
Hemisphere, the region of lowest pressure is to the left
of the direction towards which the wind blows, and, in
the Southern Hemisphere, to the right of it. The
direction of the prevailing wind, with reference to the
atmospheric pressure, is in accordance with the Law of
the Winds, as laid down by Professor C. D. Buys Ballot,
Director of the Meteorological Institute at Utrecht.
This law is summarized by Dr. Buchan as follows:—

" 1. Stand with your back to the wind, and the centre of
the depression, or the place where the barometer is lowest,
will be to your left in the Northern Hemisphere, and to your
right in the Southern Hemisphere.

" 2. Stand with the high barometer to your right, and the
low barometer to your left, and the wind will blow on your
back; these positions, in the Southern Hemisphere, being
reversed."

A third rule relates to the force of the wind, and may
be stated briefly as follows:—The strength of the wind
is generally proportional to the barometric gradient,
or the variation in the height of the barometer in
neighbouring parts.

Professor Cleveland Abbe, Meteorologist of the U. S.
Weather Bureau, makes some reference, in the recent
issue of the *Encyclopædia Britannica*, to the rotatory

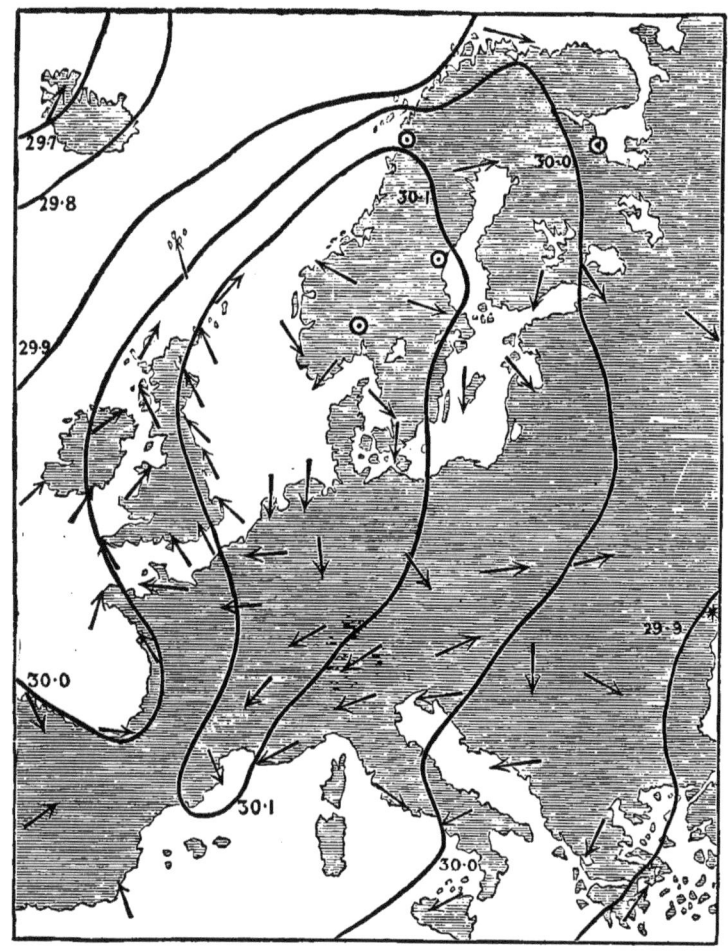

Weather Chart, showing anticyclone overspreading part of Europe,
and neighbouring cyclone.—p. 220.

(*By kind permission from the Encyclopædia Britannica*).

character of the winds, as shown in cyclonic and anti-cyclonic movements of the atmosphere. He states that the air is so extremely mobile that it quickly responds to differences in pressure which cannot be detected by ordinary barometric measurement. The pressure gradients, or barometric differences in neighbouring localities, are directly due to the fact that the atmosphere is rapidly rotating with the Earth, and, at the same time, possibly, circulating about a storm centre. No sooner is the atmosphere set in motion—possibly by inappreciable differences of pressure—than it rapidly acquires a spiral circulation. In Professor Abbe's opinion, the wind acquires this spiral circulation for two reasons :—

" (a) All straight line gusts or jets in fluids, subject to any form of resistance, necessarily break up into rotating spirals, whenever the velocity exceeds a certain limit, because the resistances of the viscosity deprive some particles of the fluid of a little more of their original velocity and energy than the other particles near by them, and thus the whole series is drawn away from linear into curvilinear paths.

" (b) In addition to their rectilinear motions, the particles of air have a rapid, circular motion, in common with the whole atmosphere, diurnally around the Earth's axis ; therefore, every particle of moving air comes under the influence of a set of forces depending on its own rate of motion relative to the Earth's surface and its position relative thereto."

The atmospheric currents, in relation to anti-cyclones and cyclones respectively, appear to be good examples of centrifugal and centripetal motion in relation to a rotating centre.

When an atmospheric movement of a cyclonic

character takes place in the Southern Hemisphere, the spiral tendency of the wind is in the direction of the motion of the hands of a watch, while in the Northern Hemisphere it is in the opposite direction. When the atmospheric movement is anti-cyclonic, the reverse is the case. This would seem to indicate that these atmospheric movements proceed in the one case—say in cyclonic disturbances—from the equatorial regions, and in the other case from the higher latitudes. An atmospheric movement of a rotatory nature, generated in the equatorial regions, would naturally have a contrary rotation, if it were propelled northward, from what it would have if propelled southward. Again, an atmospheric movement of a rotatory nature, propelled from the higher latitudes towards the equator, would naturally have a contrary rotation to a similar movement propelled in the reverse direction. There is thus a reasonable probability that the directions of revolution in these great atmospheric phenomena are mainly dependent on the Earth's axial rotation, and the place of origin of the movement.

Dr. Buchan has given some attention to the identification of the geographical places of origin and destination of the great movements of the atmosphere. In his opinion, it is evident that the regions of low and of high normal pressure, that is to say, the regions where there is usually either a comparative deficiency or a comparative superfluity of air, must be regarded as the true poles of the prevailing winds on the Earth's surface. From the unequal distribution of land and water, and their different relations to solar and terrestrial radiation, it follows that, at any rate in the Northern Hemisphere, the pole of pressure, or atmospheric movement, is very far from coinciding with the geographical pole. Dr. Buchan considers that during the winter months the regions to which the origin of the great prevailing winds of the Northern Hemisphere are to be referred are Central Asia, the region of the Rocky Mountains, and the region of calms in the Atlantic Ocean lying to the north-east of the West Indian Islands. He suggests that the regions towards and upon which the currents generated in these localities flow are the low pressure systems in the north of the Atlantic and Pacific Oceans, and the tract of low pressure within the tropics towards which the Trade Winds blow. In the summer months, when reversed conditions of pressure-distribution occur, corresponding changes in the prevailing winds take place. These facts are believed to afford the true explanation why the prevailing winds, at a large proportion of stations in the Northern Hemisphere, do not blow in the directions in which true equatorial and polar winds should blow.

We have referred to certain winds as being peculiarly dry, and to other winds as being accompanied by much rain. This, of course, has no relation to these winds considered merely as movements of the atmosphere. A wind which in one district may be absolutely parching, may, in other parts of the Earth, be usually accompanied by moisture. The difference is occasioned mainly by the nature of the surface over which the wind approaches and the geographical direction of its approach. Thus, in Britain, particularly on the west coast, the south-west wind is most frequently accompanied by rain, which is produced from the vapour which the air has taken up and carried along in its passage over the surface of the Atlantic, the direction of the approach of this wind being from a warmer region. The north-east wind, again, in the British Isles, is usually a dry wind, as it is in contact with a comparatively small area of sea before reaching our coasts, and comes from a colder region. On the other hand, in Egypt, for instance, the south-west, or southerly wind, comes from the desert, and is dry and hot; and, on the eastern seaboard of America, the easterly winds are laden with the moisture absorbed from the Atlantic.

TABLE SHOWING THE VELOCITY AND PRESSURE OF WINDS OF
VARIOUS STRENGTHS.

	Velocity in Miles per hour.	Pressure in pounds per square foot.
Calm	0	0
Light Breeze	14	1
Strong Breeze	42	9
Strong Gale	70	25
Hurricane	84	36

The broad principles which apply to the rainfall in relation to the wind may be briefly stated as follows :— If the winds have traversed a considerable extent of ocean, they bring a fair amount of rain, which is increased if their progress is towards colder regions, but is reduced or dissipated if their progress is towards warmer regions. If a range of mountains crosses the path, the rainfall is increased on the side facing the winds and reduced on the other side. If the winds, though coming from the ocean, have not traversed a great extent of it, the rainfall is not large.

Sir Archibald Geikie makes reference to the ascertained velocity and pressure exercised by the air in motion across the surface of the Earth, as being of interest from a geological point of view, and gives the above table showing the pressure according to the velocity.

Professor W. C. Unwin of the Central Technical College, Kensington, states that much attention has been given to wind action since the disaster to the Tay Bridge in 1879, and that anemometer observations show that pressures of 30 lbs. per square foot occur in storms annually in many localities, higher pressures being occasionally recorded in exposed positions. At Bidstone, Liverpool, a pressure of 80 lbs. per square foot was observed. In tornadoes, such as that at St Louis, in 1896, pressures of from 45 lbs. to 90 lbs. per square foot are estimated to have been reached. In 1881, a Committee of the Board of Trade decided that the maximum wind pressure on a vertical surface in Great Britain should be assumed, in designing structures, to be 56 lbs. per square foot.

The velocity of the wind in a tornado is usually very great. A tornado occurred in Ohio, on 4th February, 1842, which swept along at the rate of 34 miles an hour, this being the speed of travel of the tornado as a whole, as distinguished from the whirling velocity of the wind. This tornado lifted large buildings entire from their foundations, carried them a considerable distance through the air, and then dashed them to pieces. Some of the fragments were carried along for about seven or eight miles, and large oaks, nearly seven feet in girth, were snapped like reeds. A tornado passed over Mount Carmel (Illinois), on 4th June, 1877, which swept off the spire, vane, and gilded ball of the Methodist Church, and bore them bodily 15 miles to the north-eastward, thus showing the existence of an ascending current of enormous force. When a tornado passes over a dwelling house or other closed building, it often happens that the whole building is thrown outward with great violence, as if by by an explosion, proving the instantaneous and excessive reduction of atmospheric pressure outside the building. The velocity of the onward movement of the tornado has been found to vary from about 12 to about 60 miles an hour, the average being about 30 miles, and the mean time of passing a given spot is about 6½ minutes.

The hurricanes, or typhoons, of the tropics have been compared with the cyclones of higher latitudes. The differences appear to be in degree rather than in character. The tropical cyclones are of smaller dimensions, and show steeper barometric gradients—that is to say, the height of the barometer varies greatly in a limited area—and the winds are consequently stronger. The rate of progression over the Earth's surface is slower than in the case of the cyclones of higher latitudes. The nature of the atmospheric movement seems to be similar. In brief, the hurricane of the tropics would seem to be the cyclone of the temperate latitudes in a concentrated form.

It is a curious fact, that, in our observation of the wind, we almost invariably act on the assumption that its direction is horizontal. We have seen that, at the centre of a cyclonic movement of the atmosphere, there is an ascensional current of air, and that at the centre of an anti-cyclonic movement, there is a descensional current of air. In the former case air flows in from the surrounding parts at the surface ; in the latter case air flows outwards towards the surrounding parts. There seems to be strong reason to suppose that these outward and inward currents must generally have a certain angle, either of descent or ascent. In our observation of the movements of the magnetic needle, the angle of dip is considered of great importance, but in our observation of the wind this matter has been almost ignored. We are quite content to ascertain merely from what point of the compass, and with what velocity the atmospheric current flows. Yet it is not at all improbable that horizontal currents may in reality be the exception rather than the rule. In the observation of the movements of the magnetic needle also, the variation of the needle's angle to the horizontal is restricted. The needle is either horizontal or inclines earthwards. Although, north of the magnetic equator, it is the north-seeking end of the needle which inclines earthwards, and, south of the magnetic equator, the opposite end of the needle, still the angle is always a downward one. It is invariably an angle of *dip*. In the observation of the wind a wider range must be given. The wind may either be horizontal or, on the one hand, it may have any angle of descent up to—in the case of a descending current—90 degrees; or, on the other hand, it may have any angle of ascent up to —in the case of an ascending current—90 degrees.

Dr. Buchan mentions that "observations of the winds cannot be conducted, and the results discussed, on the supposition that the general movement of the winds felt on the Earth's surface is horizontal, it being evident that the circulation of the atmosphere is affected largely through systems of ascending and descending currents." Notwithstanding this, observations generally, respecting the wind, are still mainly confined to its horizontal direction and its velocity.

Let us now consider the cyclonic and anti-cyclonic movements with reference to the question of the deviation from the horizontal of the surrounding air-currents. The cyclonic movement has, at its centre, an ascending column of air. Now, this central movement may itself be either vertical or may deviate somewhat from the vertical. As the movement of the air cannot itself be observed, information as to its character is obtained from the observation of clouds and dust storms.

Dr. Buchan states that in a tornado the usual position of the gyrating columns of cloud is vertical, but, occasionally, a curving form or slanting direction is observed. He suggests that, to these latter forms, many stationary or slowly moving, dangerous squalls, which spring up suddenly in lakes and arms of the sea in mountainous regions, may owe their origin. In the case of dust storms, a similar state of matters appears to exist. The simplest form of the dust storm, as it occurs in India, Arabia, or Africa, has a tall aerial column of sand moving onwards, and drawing into itself as it whirls along, dust and other light bodies. These are swept inwards by the strong air-currents which blow along the surface of the ground, and converge with a gyratory motion towards the base of the column. A common form of dust storm has several dust columns grouped together, each whirling independently on its own axis, and the whole group gyrating as it is borne onward; and the forms and relative positions of the columns changing. The observation of dust storms shows that there is a strong inflow of the air along the surface of the ground all round vorticosely towards the base, and that these inflowing air-currents afterwards ascend along the central axis of the column, carrying

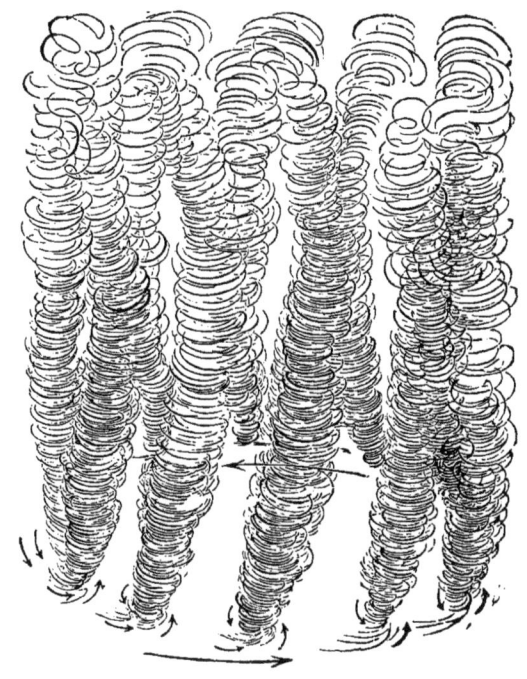

Dust Storm.—p. 228.

(By kind permission from the Encyclopædia Britannica).

with them particles of dust, sand, and other light objects. It seems probable that the atmospheric movements revealed by these storms are typical, in an exaggerated form, of the ordinary cyclonic movement of the atmosphere. Thus there seems to be some reason to suppose that the central air movement in either a cyclonic, or an anti-cyclonic, area may be either vertical, or have a curving form or slanting direction.

If, now, we have a region where the air has a somewhat vertical movement of ascent, how will this affect the movement of the surrounding air as regards deviation from the horizontal? It would seem probable that the inflowing air from the surrounding parts will necessarily have a movement of descent. The purpose of the inflow is to supply the deficiency of air produced at the surface by the ascending movement, and, in order to supply this deficiency, there must be a rush of air from all the surrounding parts. As the deficiency occurs on the surface there can be no inrush from *lower* levels of the atmosphere. The rush must be from either the same level or higher levels, and, as all the neighbouring air will be equally impelled, it would seem that this would result in the general flow of the air having a movement of descent.

Exactly the converse would seem to be the case in the anti-cyclonic movement. There the central movement is one of descent. A great column of air is flowing downwards, striking the surface and out-flowing in all directions. The outward rush must be either horizontal or upwards, and it will most likely partake of both characters; so that, as regards deviation from the horizontal, the movement will be ascensional.

According to this reasoning, the direction of the wind around the centre of a cyclonic area will be inwards and downwards, and around the centre of an anti-cyclonic area outwards and upwards, the general deviation from the horizontal probably being but slight.

It is very remarkable that, according to these suppositions, the inclination of the wind, immediately outside the centre of either a cyclonic or an anti-cyclonic system has the character which, when carried to its extreme, marks the centre of the opposite system. The cyclonic system, at the outer margin of its central area, will be characterized by a descending air-current, and this, when it approaches the vertical, is the characteristic of the centre of an anti-cyclonic system. The anti-cyclonic system, at the outer margin of its central area will be characterized by an ascending air-current, and this, when it approaches the vertical, is the characteristic of the centre of a cyclonic system.

It would seem likely that, if these suppositions are correct, the chief departure of the air-currents from the horizontal in each system (if the central areas with their vertical, or almost vertical currents, are left out of account) will be in the *immediate neighbourhood* of the central area. As the distance from the centre is extended the deviation of the wind from the horizontal will become less marked, until, at length, a distance is reached when there is no such departure. This may be supposed to mark the boundary of the system; or, probably—if the anti-cyclone may be taken to be the complement of the cyclone—the boundary of two opposite systems. Where the anti-cyclonic system ends the cyclonic probably begins. If this is so, then, when we reach the position at the bounds of an anti-cyclone, where the wind has no deviation from the horizontal, we shall, by continuing in the same direction, gradually reach a locality where the wind has a downward movement, which will indicate that we have entered upon a cyclonic area. The converse would occur if the observation were commenced immediately outside the central area of a cyclonic system.

Of course, in observations of the angle of the wind in relation to the Earth's surface, the physical characteristics of the surface in the district would have to be considered. It is evident that, in mountainous regions, the direction of the wind, especially in the lower levels, may be subject to exceptional causes. The most suitable localities for observation would be those which were free from obstacles to the natural course of the wind.

The whole subject of the deviation of the wind from the horizontal is very vague and uncertain. It is, in fact, a new field for meteorological observation. The subject appears to be of importance, and its patient investigation should prove advantageous to meteorological science, and to our knowledge of weather conditions.

There does not seem to be yet in existence (1906) any instrument by which the ascensional or descensional angle of the wind can be accurately ascertained. The subjects of present observation in relation to the wind are its direction, and its velocity or pressure. The pressure varies with the nature and method of exposure of an obstacle to the wind's progress, and is not considered by meteorologists of such importance as the velocity. The instrument used for measuring the velocity is called an anemometer. In its best known form this instrument consists of four hemispherical cups, which are caused by the wind to revolve horizontally around a vertical axis. The speed of revolution, which is registered by a mechanical arrangement, indicates the velocity of the wind; but the relation varies with the dimensions of the cups and arms, and the steadiness or gustiness of the wind. The results arrived at also vary according to the position and height of the instrument. The cup-anemometer has been largely superseded by the vertical suction-tube anemometer, which is associated in Britain with the name of Mr. W. H. Dines. Two vertical tubes, whose apertures are respectively directed to windward and leeward, and within which are two independent barometers, give the means of determining the barometric pressure, either *plus* or *minus* the wind pressure, so that both the velocity of the wind and the true barometric pressure can be determined.

No uniform rules as to the exposure of the anemometer have yet been adopted, and, for the purposes of meteorology, the comparative records of the anemometer cannot be considered as altogether satisfactory. Apparently, no form of this instrument is adapted for showing the angle of inclination of the wind.

The study of the winds is really a branch of the study of the atmosphere. We have seen that the atmosphere is kept in continuous agitation through the rotation of the Earth, and the unequal distribution of heat and cold, and that the currents, which are thereby set up, are what we, who live at the bottom of the atmospheric ocean, describe as winds. As, in their origin, these currents have a movement of ascent or descent, it seems unreasonable to endeavour to arrive at a complete

Cup Anemometer. —p. 232.

(*By kind permission of Messrs Negretti and Zambra*).

understanding of the laws by which they are controlled by proceeding on the assumption that they flow along the surface of the Earth. We are thereby leaving out of account certain of the main factors in the problem. The value of these disregarded factors it is impossible to estimate. They may, or they may not, be of great value. It is clear, however, that, in the investigation of the laws of the atmospheric currents, they should not be ignored. It is possible that their due recognition may prove of the utmost importance in the elucidation of the great problem of the laws of the winds, and in the forecasting of changes in the weather—changes which are uniformly coincident with variations in either the direction or the velocity of the wind.

THE DAY AND THE PLACE
OF ITS BIRTH.

SYNOPSIS.

Local differences in time according to longitude—Diversity of time in solar day as compared with civil day—Variation in the length of the solar day—The causes of the variation—Longest days actually occur in the end of December—Cumulative effect of variation in length of solar days—Solar days of nearest approach to 24 hours in length—Solar days on which the Sun crosses the meridian at 12 noon of mean time—Specialities of the Arctic and Antarctic regions—Cardinal direction of the Sun at solar noon in different geographical positions—Equal duration of sunshine all over the Earth—Mean length of solar day—Solar time-interval between different geographical positions—Impracticability of fixing local civil day by local solar time—Day lost or gained in journeying round the world according to direction—Variation in length of solar day from artificial causes—The date line—Historical and geographical causes affecting its course—Standard time—The zone system—Modes of adoption of standard time—The application of the zone system to Ireland—Unique characteristics of region of the date line—The " Earth day."

THE DAY AND THE PLACE
OF ITS BIRTH.

IF, at the hour of noon in Greenwich time, we could be instantaneously transferred from London to Paris, we should find that the time was nine minutes past noon; and, if we could be similarly transferred to Dublin, we should find that we had gained on the clock by 25 minutes —that the time was only 11.35 a.m. This, of course, arises from the difference of the longitude of Paris and Dublin as compared with that of Greenwich.

In its apparent daily westward circuit of the heavens, the Sun, generally speaking, decides the hour of the day. It exactly decides the hour of the true, or solar, day, and, subject to a trifling correction of a variable character, it decides the hour of the civil day. Thus, when the Sun is to the east of the local meridian, it is invariably fore-noon in the solar day; when it is on the meridian it is noon; and when it is to the west of the meridian it is afternoon. This is indicated by the terms *ante meridian* and *post meridian,* which we constantly use in their abbreviated forms of a.m. and p.m. These terms, however, apply literally only to the solar day, and not to the civil day.

The length of the solar day is not constant, but varies slightly throughout the year. There are two causes of this variation, the first being the mutability of the distance of the Earth from the Sun, and the second, the apparent northward and southward movement of the Sun throughout the changing seasons. Let us consider these causes in some detail, and the manner in which they affect the length of the solar day.

The distance which separates the Earth from the Sun changes day by day, during the progress of the year. The average distance is about 92,897,000 miles; but, owing to the elliptical shape of the Earth's orbit, the distance ranges from about 94,447,000 miles when it is greatest, to about 91,347,000 miles when it is least. The difference between the greatest and least distance of separation is thus more than three millions of miles. The speed of the Earth's movement in its orbit varies according to the distance which separates the Earth from the Sun. When the Earth is nearest to the Sun its orbital velocity is greatest; when it is farthest away from the Sun its orbital velocity is least. The change in the velocity is thus inverse to the change in the distance, —the velocity increasing as the distance becomes less, and decreasing as the distance becomes more. As the Earth's orbit is a circuit around the Sun, approximating in shape to a circle, it is evident that the Earth, in its orbital journey, must pass over an arc of this circuit in each day's progress. The length of the arc which is passed over will vary with the speed at which the Earth is moving.

Now, the time which the Earth takes to rotate on its axis does not vary. It is always, throughout the year, 23 hours, 56 minutes, 4 seconds. After an interval of that duration each meridian is again turned towards exactly the same part of the heavens as at the beginning. The meridian, however, which at the beginning of that time was turned directly towards the *Sun*, is not turned directly towards it at the end of that time.   This arises from the fact that, in the interval, the Earth has moved a certain distance around the Sun.   The meridian, therefore, which faced the Sun at the beginning has to be turned a little further round than it would be brought merely by the completed axial rotation, in order to bring it again to face the Sun.   When the axial rotation is completed, and the meridian again faces the same part of the heavens, the Sun is no longer there.   It appears as if the Sun had moved a little to the east in the intervening time, the cause of this being that the Earth has moved a little in the opposite direction.

It is manifest, that the longer the distance in its orbit which the Earth has passed over in the interval, the further will the meridian have to be turned round, after the completion of the axial rotation, to bring it again to face the Sun.   The less the distance in its orbit which the Earth has passed over, the less will the meridian have to be turned round in order to face the Sun.   As the Earth must make *some* appreciable progress in its orbit in the course of 24 hours, even when its orbital velocity is smallest, it is evident that the meridian will always have to be turned round a little, after the completion of the axial rotation, before it will again face the Sun.   The axial rotation occupies, as we have noticed, 23 hours, 56 minutes, 4 seconds, this period being known as the sidereal day.   As the solar day is the interval which elapses between the time when a meridian faces the Sun and the time at which the same meridian decreasing as the distance becomes less, and ... distance becomes more.   As the **Earth's** ... around the Sun, approximating in shape ... evident that the Earth, in its orbital ... over an arc of this circuit in ... The length of the arrival ... into the speed at ...

exactly the same part of the heavens as at the beginning. The meridian, however, which at the beginning of that time was turned directly towards the *Sun*, is not turned directly towards it at the end of that time. This arises from the fact that, in the interval the Earth has moved a certain distance around the Sun. The meridian, therefore, which faced the Sun at the beginning has to be turned a little further round than it would be brought merely by the completed axial rotation, in order to bring it again to face the Sun. When the axial rotation is completed, and the meridian again faces the same part of the heavens, the Sun is no longer there. It appears as if the Sun had moved a little to the east in the intervening time, the cause of this being that the Earth has moved a little in the opposite direction.

It is manifest, that the longer to distance in its orbit which the Earth has passed over in the interval, the further will the meridian have to be turned round, after the completion of the axial rotation, to bring it again to face the Sun. The less the distance in its orbit which the Earth has passed over, the less will the meridian have to be turned round in order to face the Sun. the Earth must make *some* appreciable progress orbit in the course of 24 hours even when velocity is smallest, it is evident that the always have to be turned round a now west- cir- as the onds—

again faces the Sun, it is plain that even the shortest solar day must be longer than the sidereal day. It is also evident that, as the extent which a meridian has to be turned round in order to face the Sun, after the completion of the Earth's axial rotation, must vary with the distance which the Earth has travelled in its orbit in the interval, the length of the solar day must vary according to the orbital velocity of the Earth. When the orbital velocity is greatest the length of the solar days will be greatest, and when the orbital velocity is smallest the length of the solar days will be shortest. The Earth is nearest to the Sun in the beginning of January, and farthest away in the beginning of July. This cause of the variation in the length of the solar days would, therefore—if it alone operated—give us the longest solar days at the former season, and the shortest at the latter. This cause, however, does not operate alone, and we shall now consider the second cause of the variation and its effects.

The length of the solar day is, as we have mentioned, influenced by the apparent northward and southward movement of the Sun throughout the changing seasons. This apparent movement results from the plane of the Earth's orbit not being the same as that of either the terrestrial equator or any parallel of latitude—that is, not being at right angles to the Earth's axis. If the plane of the orbit were at right angles to the Earth's axis, the Sun would, during the whole year, be in the zenith either on the same parallel of latitude or on the equator, and the seasons would not exist. The Earth, however, in passing around the Sun, moves at one part of its orbit somewhat in the direction which, in relation to the terrestrial surface, we consider "northward"; while, at the opposite part of the orbit, it moves somewhat in a "southward" direction. These orbital movements of the Earth, "northward" and "southward," cause an apparent movement of the Sun in the opposite direction. Let us now leave out of account the difference which occurs in the distance separating the Earth and the Sun, and assume that it is constantly the same, and see how this "northward" and "southward" movement of the Earth will affect the length of the solar day.

Evidently, if the orbit of the Earth were on the same plane as the equator, or of any parallel of latitude, and the distance of the Earth from the Sun were constantly equal to the present *average* distance, the length of the solar day would not vary. It would always be of the same length as the present civil or mean day, that is, exactly twenty-four hours. If again, the orbit of the Earth were on the same plane as one of the terrestrial meridians—or, perhaps, we should say, as two opposite meridians, for instance the meridian of Greenwich and the 180th meridian—then also (supposing the distance from the Sun to be constant) the length of the solar day would not vary. In this case, the apparent *annual* movement of the Sun, instead of being eastward in the heavens as at present, would be due north and due south, while (as we are not supposing the nature of the axial rotation to be different in any respect from what it now is) the Sun's apparent *daily* movement would be westward as at present. The solar day would, in these circumstances, be invariably of the same length as the sidereal day—that is 23 hours, 56 minutes, 4 seconds—.as the movement of the Earth in its orbit would be at right angles to the daily movement of rotation. Now, the actual orbital movement of the Earth is of a nature between these two supposititious forms of movement. The orbit is not on the plane of either a parallel of latitude or a meridian of longitude, but its plane is between these, and forms an angle with the equator of about 23½ degrees.

If, now, we suppose the Earth's orbit to be represented by a sloping circle on the plane of this paper, like the capital letter *O* tilted to one side, we shall have a simple representation of the effect which the inclination of the orbit has on the length of the solar day. A line passing through the centre of this tilted letter from left to right, parallel with the printed lines, would, we shall suppose, be on the plane of the equator. It is evident that, near this central line, the slope upwards and downwards is much more marked than it is at the curved extremities of the letter. Thus—to apply our illustration— immediately before and after the equinoxes, or the time when the Sun is on the equator, the orbital movement of the Earth tends much more noticeably in a direction oblique to the plane of the equator, or, as we may express it, more "northward" or "southward" (and consequently the apparent movement of the Sun is much more markedly southward or northward), than it is at and near the summer and winter solstices. At the solstices, the extreme "northward" or "southward" position in the orbit is reached, and the course changes. Therefore, at and near the equinoxes, the Earth in a day's orbital journey, will make considerably less *"forward"* progress or advance in its orbit, as measured on the plane of the equator (or any plane parallel

thereto), than at and near the solstices. At the former
period, a great part of the progress is not "forward,"
that is to say, is not on the plane of the equator or any
parallel of latitude, but is "northward" or "southward;"
while, at the latter period, the orbital movement is
almost . entirely a "forward" one. As the apparent
annual eastward circuit of the Sun is occasioned by the
actual orbital movement of the Earth, the Sun will, at
the equinoxes, appear to have moved a less distance
towards the east after an interval of 24 hours than it

DIAGRAM ILLUSTRATIVE OF THE EARTH'S ORBITAL MOVEMENT.

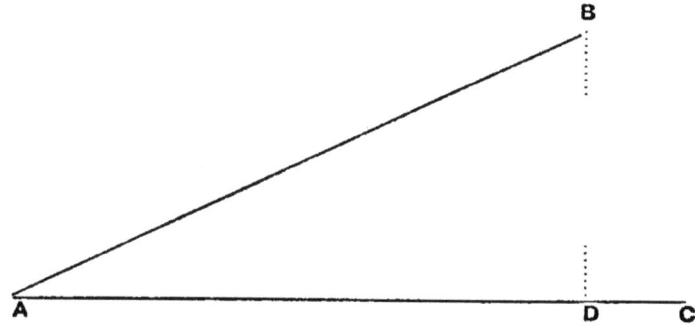

B and C are two points starting simultaneously from A, with equal
velocities, to the margin of the paper, the one diagonally upwards, the other
directly. It is evident that, owing to its upward motion, B will take longer
to reach the margin than A, the *direct* velocities to the margin for a given
period being represented by the lines AD and AC respectively.

will appear to do at the solstices. At the former
seasons, the apparent movement will be of a somewhat
north-easterly or south-easterly character, while, at the
latter, it is more directly eastward.

At the solstices, the orbital movement of the Earth
resembles the progress of a person walking along a level
road, while at the equinoxes it resembles the progress
of a person walking diagonally up a hillside. Supposing
the *lineal distance* passed over in a given time by two

pedestrians, one on the level and the other mounting upwards, to be exactly the same, it is, nevertheless, evident from the diagram (see p. 243) that the person walking on the level will, at the end of the time, be a greater distance from the starting point than the person moving upwards, *as regards the distance travelled in a line parallel to the level.* In the case of the Earth what we may call *the change of level* does not alter the interval between succeeding returns of the meridian to the Sun. This interval is affected only by the " forward " movement. As we have seen, if the orbital movement resulted merely in a change of level, the solar day would correspond with the sidereal day. The same meridian would face the Sun every 23 hours, 56 minutes, 4 seconds, although the Sun would be in the zenith at a different part of it,—that is to say, the Sun would appear to be either higher or lower in the heavens than when previously on the meridian. It is, therefore, clear that the length of the solar day will vary according to the degree of prominence in the orbital movement of the Earth of a} "northward" or "southward" change of position. Supposing the distance of the Earth from the Sun to be constant, then, at such time as the "northward" or "southward" change of position is most prominent in the orbital movement, the solar day will be shortest; when there is least "northward" or "southward" movement, the solar day will be longest. In the first case the meridian, after the completion of the diurnal rotation, has not to be turned so far backwards, in order again to face the Sun as it has in the latter case. Thus, this cause of variation will tend to give us the shortest solar days at and near the equinoxes, that is about 21st

March and 23rd September, and the longest solar days at and near the solstices, that is about 21st June and 22nd December.

Either of these causes of variation would manifestly by itself alone produce a constant change throughout the year in the length of the solar day, and, as they both exist, they will exercise a modifying influence on one another. The variation in the distance separating the Earth from the Sun would result in the longest solar days occurring in the end of December and the beginning of January, and the shortest in the end of June and the beginning of July, while the apparent northward and southward movement of the Sun during the year—that is, the obliquity of the ecliptic—would result in the longest solar days occurring at the solstices in June and December, and the shortest at the equinoxes in March and September. Both causes, therefore, act in concurrence in the end of December, while they act in opposition in the end of June. It thus comes about that the closing days of the year, which we, in these northern latitudes, regard as the shortest "days," are, if considered as solar days, not only not the shortest, but are actually the longest days in the whole year. Of course, the excess length is really inconsiderable, but, still, for some weeks in succession at this season, each solar day is about half a minute longer than the mean length of 24 hours. In the end of June and beginning of July, when the causes of variation act in opposition, the effect, in lengthening the solar day, of the greater orbital progress or "forward" movement (as distinguished from the "northward" or "southward" movement), is greater than the effect of the increased distance from the Sun in reducing it, and, thus, the solar day is then about 12 seconds longer than the average. At the equinoxes, again, the joint effect is to reduce the length of the solar day, so that it is then from 18 seconds to 20 seconds shorter than the mean length. The varying correction which has to be made to bring solar time and mean time into harmony is known as the "equation of time."

We must keep in mind that the slight differences which exist between the length of the solar day and the civil day are continued in the same direction for weeks at a time; so that although, in one day, the difference is, when greatest, only about half a minute, it ultimately, by accumulation, becomes of some consequence. Thus, by about the 11th of February, the accumulated effect of the greater length of the solar days places the Sun, when it reaches the meridian, nearly a quarter of an hour (14 minutes, 26 seconds) behind the clock; while, in the beginning of November, the accumulated effect of the shorter solar days enables the Sun to reach the meridian about 16 minutes, 20 seconds, before the clock shows the hour of noon. Although the mean length of the solar day is exactly 24 hours, there are very few, if any, solar days which are precisely this length. The nearest approach to it occurs at the time of maximum variation between the Sun and the clock, when the difference, after having slowly accumulated, begins at last to decrease. This happens about 11th February, 15th May, 27th July, and 3rd November.

In view of the constant variation in the length of the solar day, it is scarcely surprising to find that, although when the Sun is on the meridian it is necessarily exactly 12 o'clock in the solar day, this seldom, if ever, happens at exactly 12 o'clock in the civil day.   It, in fact, occurs with exactitude, or nearly so, on only four days in the year—the dates when it happens being on, or about, 15th April, 14th June, 31st August, and 24th December.   Such harmony between the Sun and the clock as regards the hour of noon, is, however, always coincident with a very appreciable difference between the lengths of the solar and civil days.

It should be mentioned that, within the Arctic and Antarctic circles, in the summers of these regions the Sun may be twice on the meridian during the solar day, throughout a period varying from one day to six months, according to latitude.   On both occasions, however, this is at exactly 12 o'clock (solar time), the first occasion being at 12 o'clock noon and the second at 12 o'clock midnight. Within the Arctic circle the direction of the Sun, when visible at noon, is due south, while, when visible at midnight, it is due north, the converse being the case within the Antarctic circle.   The double crossing of the meridian in the solar day during the summer season is, of course, compensated for by the absence of the Sun for an equal period during the winter season. In all other parts of the Earth the Sun is only once on the meridian during the solar day.   Its position at noon in the temperate latitudes of the Northern Hemisphere is due south, while in the corresponding latitudes of the Southern Hemisphere it is due north. Within the tropics the direction of the Sun at noon varies according to the season and the latitude.   At the tropic of Capricorn, which is about $23\frac{1}{2}$ degrees south of the equator, the Sun is in the zenith at noon at the December solstice ; at other seasons, due north of the zenith at the same hour. At the tropic of Cancer, which is an equal distance north of the equator, it is in the zenith at noon at the June solstice; at other seasons, at the same hour, it is due south of the zenith. At the equator it is in the zenith at noon at the equinoxes, while, at the same hour, between the vernal and autumnal equinox it is due north of the zenith, and between the autumnal and vernal equinox due south, of the same point, although its elevation is always very great. It is an interesting fact that, whatever the latitude, every part of the surface of the Globe has the Sun above the horizon for one half of the year, and below it for the remaining half. Indeed, the time during which the Sun is above the horizon in the course of a year slightly exceeds the time it is below the horizon, whatever the geographical position. The excess is occasioned partly by the size of the Sun in comparison with the Earth—whereby slightly more than half the Earth is, at the same time, directly exposed to the Sun's rays—but, chiefly, by the refraction of the rays produced by the atmosphere, whereby the Sun continues visible for a short time after it has actually set.

In the course of a year, the accumulated time by which solar days exceed 24 hours is exactly equivalent to the time by which other solar days are less than 24 hours, so that the mean length of the solar day is precisely 24 hours. This period has, therefore, been fixed as the invariable length of the civil day, which it would, of course, be utterly impracticable to have of a variable character.

We have now seen that (subject to the qualification specified in regard to the regions within the Arctic and Antarctic circles respectively), a period of, on the average, 24 hours intervenes between successive returns, through the Earth's axial rotation, of the same meridian to the Sun. That is to say, the Sun, apparently passes over the 360 degrees forming the circumference of the Earth, from any meridian to the same meridian again, in a period of, on the average, 24 hours. This makes the mean distance in longitude, which the Sun apparently travels in its hourly journey, 15 degrees. It is, therefore, very easy to calculate the solar time-interval between one place and another. We have merely to ascertain their relative positions in longitude, and work out the time-interval on the basis of 15 degrees of longitude representing one hour in time. As the Earth's rotational movement is from west to east, whereby the Sun apparently moves from east to west, it is evident that in the direction from which the Sun appears to come—that is the east—the hour of the day must be later than it is with us; and that in the direction towards which the Sun appears to be travelling—the west—the hour of the day must be earlier than it is with us. The Sun may be said to carry noon with it. When, therefore, the Sun reaches us we have noon; while noon has already passed, with the Sun, from the parts to the east of us; and has still to be carried by the Sun to the parts to the west of us.

We have noticed that, when it is noon in London, it is said to be nine minutes past noon in Paris, and twenty-five minutes to noon in Dublin. We can readily test this statement in the method specified. The time of noon in London is fixed by the mean time of the Sun's passage over Greenwich Observatory, and that of Paris and Dublin by its passage over the local Observatories. Paris

Observatory is 2° 20' 14" E., and Dublin (or Dunsink) Observatory is 6° 20' 15" W., of that of Greenwich. As 15° represents one hour, 2° 20' 14" will represent 9 minutes, 20·9 seconds, and 6° 20' 15" will represent 25 minutes, 21·1 seconds. Thus, at noon in London the exact time in Paris is 12 hours, 9 minutes, 20·9 seconds, and in Dublin 11 hours, 34 minutes, 38·9 seconds; but, of course, such exactitude is not practicable in real life, and the times are accordingly taken, with quite sufficient accuracy, as 12.9 p.m. and 11.35 a.m. respectively. It will be noticed that the times mentioned are not strictly applicable to either London, Paris, or Dublin, but to the Observatories by which the clocks in these cities are respectively regulated.

It is, indeed, obvious that it would be quite impracticable to have the civil day in each separate locality fixed by the mean local solar time. In London, for instance, there is an appreciable difference between the mean solar time of the east end and the west end. If we take the longitudinal stretch of London as 14 miles, and take the length of the degree of longitude in the latitude of London (say, 51° 30' N.) as 43¾ miles, then the longitude of London will cover 19' 12", which would represent a difference between the mean solar times of the east and west ends of more than a minute and a quarter. The absurdity of any attempted recognition, in our time-reckoning, of such a difference is evident. It is, therefore, a matter of necessity, almost as much as a matter of convenience, that the time of certain extensive and easily defined areas should be regulated by the mean time of the transit of the Sun over a certain recognized position within the area. Accordingly, the time, for the whole of Britain, is fixed by the mean time of the Sun's passage over the meridian of Greenwich, or rather by the mean time at which that meridian is, by the Earth's rotation, caused to face the Sun. The times for France and for Ireland are, in like manner, regulated by the Observatories at Paris and Dublin respectively, and a similar state of matters prevails in almost all other countries.

In view of the fact that to the east the time is later, and to the west earlier, than it is on our own meridian, a curious and interesting state of matters—which has been skilfully made use of in certain romances—results from continuous progress in either direction. Let us suppose that in London it is noon on Monday, and let us imagine ourselves to pass round the world in an easterly direction with the speed of the electric telegraph, or, say, the speed of light. At a distance of 15 degrees to the east of London we should find that it was one o'clock in the afternoon; at 30 degrees it would be two o'clock, and so on. At the 180th meridian it would be twelve hours later than in London, so that the time would be twelve o'clock midnight between Monday and Tuesday. Keeping a straight course, we complete the circuit of 360 degrees in a moment of time, and are back again in London. It would, however—according to the rule that for every 15 degrees of eastward movement the time is one hour later—be, apparently, noon on Tuesday. An equally absurd result is arrived at by imagining the accomplishment of an instantaneous westward circuit. In this case we have to take off one hour for every 15 degrees of westward progress, so that, after an instantaneous journey of 360 degrees westward, we should arrive back in London at noon on the day preceding our setting out. This is, of course, the *reductio ad absurdum* of what is an actual physical fact, which is, that in journeying round the world in an easterly direction, a day is lost, while, if the journey is done in a westerly direction, a day is gained.

Let us suppose that two ships leave Southampton on the same day for a voyage round the world, the one travelling eastward the other westward. Let us assume that both vessels arrive back on the same day after completing their respective voyages. Notwithstanding the times of the departure and arrival being the same, the vessel which travelled westward would have accomplished the voyage in two days less time than the other. Supposing that some of the sailors in these vessels had kept a record of the passing days, they would find, on comparing notes with the landsmen, that some confusion existed as to the day of the week. We may imagine that the landsmen believed the day of the arrival of the vessels to be Tuesday. The sailors who had travelled eastward would have good reason to believe it to be Wednesday, while the crew of the other vessel would have equal confidence in considering it to be Monday.

The explanation of these facts is, simply, that in travelling eastward we are travelling with the Earth's rotation and against the Sun, while in travelling westward we are travelling against the rotation and with the Sun. While the mean length of the solar day is 24 hours, it is evident that by travelling eastward we are passing towards other meridians which will face the Sun earlier than the meridian which we are leaving, so that between noon one day and noon the next day, will, in our experience, be less than the usual mean interval of 24 hours. In travelling westward, we are passing towards other meridians which will face the Sun later than the meridian which we are leaving, so that between noon one day and noon the next day, will, in our experience, be longer than the usual mean interval of 24 hours. There is no possible doubt that the absolute time which intervened between the departure and the return of the vessels, in their supposed voyages, was of the same duration. We may suppose the time to have been a year—365 days. The length of these days, to the landsmen in Southampton, was evidently 24 hours each; but, with the voyagers, this was not the case. Those who travelled eastward had passed through 366 days, of the average length of about 23 hours, 56 minutes, 4 seconds; while those who travelled westward had passed through 364 days, having an average length of about 24 hours, 3 minutes, 57 seconds. The day gained in the one case was exactly compensated for by the shorter length of the days, and the day lost in the other case was exactly compensated for by the greater length of the days.

It is thus clear that, while the mean length of the solar day is 24 hours, the actual length varies not only from natural causes, but, in personal experience, from artificial causes. It is quite possible to make the length of a day in our personal experience 25 hours instead of 24 hours, or, on the other hand, to reduce it to 23 hours. All that requires to be done to shorten the day to 23 hours is to pass it in travelling eastward, so as to cover 15 degrees of longitude. In the latitude of London, the necessary distance would be about 650 miles, and, as the length of the degree of latitude becomes less as the latitude increases, the distance which would have to be passed over would be less, farther north than London. A similar journey in the opposite direction would correspondingly increase the length of the solar day.

It would certainly be very inconvenient to have confusion arising as to the day of the week, such as we have seen would occur to travellers round the world were no remedy artificially provided. This must have become evident very soon after such journeys began to be made, and the means of overcoming the difficulty would soon be apparent. The remedy is, merely, that those who travel round the world in a westerly direction should omit a day in their reckoning, and that those who accomplish the same journey in the opposite direction should, in their calendar, repeat one day. Evidently, in order to allow of this remedy being applied practically, an international arrangement should be come to as to the part of the Earth at which the necessary correction should be made. Such an arrangement cannot be said to be yet finally completed, but, by gradual agreement, the 180th meridian has, as nearly as possible, come to be recognized as the place at which the change should be made—as the part of the Earth at which the day may be considered as having its birth. This arrangement has been materially helped forward by the fact that, almost throughout its whole course, the 180th meridian extends over the sea. It is evident that it would be very inconvenient to have a change of day at different sides of an imaginary line extending to any great extent over the land surface, especially in populous countries; and, where the required correction is applied, the day must necessarily be different on the two sides of this imaginary line, as the time-difference is 24 hours.

It is, no doubt, owing to the British command of the seas, and to the predominance of British commerce, that the place where the day changes has slowly come to be recognized universally as being mainly the 180th meridian. This meridian is, of course, exactly opposite the meridian of Greenwich, and it and the meridian of Greenwich form together a great circle around the Earth, extending through both Poles. It is natural that, when it became evident that such a correction as this required to be made when travelling round the world, the early British circumnavigators should make the correction at the greatest possible distance in longitude from the meridian which regulates the time in Britain, and, no doubt, their early practice to this effect has had a decisive influence with subsequent travellers of other nations.

It is only, however, very gradually that the 180th meridian is becoming generally recognized as the place where the day changes, and, even still, it does not control the day throughout its whole course. This incompleteness might reasonably be expected where the meridian intersects inhabited lands, on account of the public inconvenience which would result from people living in the same district, and separated only by an imaginary line, keeping the passing day as two distinct days of the week—as two distinct dates of the month. Where the meridian passes through inhabited land, therefore, the tendency has been to accept the nearest sea channel as forming the dividing line. Just indeed, as different nations, having dominion over conterminous lands, endeavour to adjust a convenient and easily recognizable boundary, so the different nations have, in this matter, to adjust a convenient and easily recognizable boundary between one day and another.

Up to about the middle of the nineteenth century there was very great irregularity in regard to this line of demarcation, or, as it is called, the "date line." The natives of the Philippines, although the Islands lie far on the other side of the 180th meridian from that on which America lies, kept the date which applied to America. This custom had its origin in the fact that the Spaniards originally approached these Islands from the Pacific coast of America, and, as might be expected, they carried the date with them. In these early days of navigation there would naturally appear to be no necessity for making a change of date in a journey, which, in comparison with the circumference of the Earth, was of small account. Thus Luzon, in the Philippine Islands, and Celebes, in the East Indian Islands, although practically on the same meridian (say about 120° E. long.) kept different dates. This was remedied by the Archbishop of Manila, who had jurisdiction over the Philippine Archipelago. He decreed that the 30th of December, 1844, should be immediately followed by the 1st of January, 1845, a day being thus dropped, and the date applicable to the Philippines brought into conformity with that applicable to other places in the same longitudes. This had a very considerable effect in the straightening of the date line.

Another matter which gave progress in the same direction was the purchase of Alaska from Russia by the United States, in 1867. This great territory lies on the United States side of the 180th meridian, its most western extremity not extending so far west as the 170th meridian west longitude. Notwithstanding this, when it belonged to Russia, its day was, as we should surmise, the same as that kept in Siberia on the other side of the 180th meridian, and was, therefore, different from that kept in the Canadian territory, to which it is geographically united. The United States, on acquiring the country, changed the day into harmony with that kept in the rest of America.

Dr. A. M. W. Downing, F.R.S., has given some attention to this subject, and he expresses the opinion that "further changes in the direction of the assimilation of the date line to the 180th meridian would, at present, be difficult, as its course is mainly determined by the grouping of the islands, and by the particular circumstances in each group upon which depends the direction in which it has intercourse with the outer world." He considers that "these circumstances may alter as time goes on, and may be followed by corresponding changes in the position of the date line, but the assimilation, if carried out at all, must be a gradual process."

The deviations which the date line still makes from the 180th meridian arise from various causes. If the line conformed to the meridian it would cut off a peninsula of considerable extent from Siberia—the East Cape—and give to the inhabitants of that district a different day from that kept in the neighbouring land under the same rule. This would necessarily cause inconvenience, so that the deflection, which has to be made to prevent it, may be said to be unavoidable. There is the less reason to regret this deflection, as it gives us a most convenient course at no great distance, which lies right through Bering Strait, which in this part separates Asia from America.

The remaining causes of deflection of the line are more regrettable. An extensive incursion has to be made in the direction of Asia to include the long archipelago constituting the Aleutian Islands, so as to give these islands the American day. This chain of islands was included in the property purchased by the United States from Russia in 1867, when Alaska was acquired. Prior to their transference, the islands, like Alaska, had the Asiatic day, and, of course, when they were acquired by the States the day was changed. The greater part of the chain certainly lies on the American side of the 180th meridian, but, unfortunately, the chain extends about 8 degrees beyond the meridian in the direction of Asia. The islands number about 150 in all, and have about 8,000 inhabitants. They are of small importance, but it would certainly be rather awkward to have the day different in some of them from the rest. It would be a bold policy to carry out. What makes this deflection the more noticeable is that it occurs close to the equally large deflection in the opposite direction caused by the Siberian peninsula of East Cape.

After it passes the Aleutian Islands, in its southward course, the date line conforms to the 180th meridian until it reaches the islands of the South Pacific, that is, from about 49° N. latitude to, perhaps, about 12° S. latitude. A long easterly deflection is then caused by the line having to take the eastern side of the Friendly Islands, the Kermadec Islands, and the Chatham Islands,

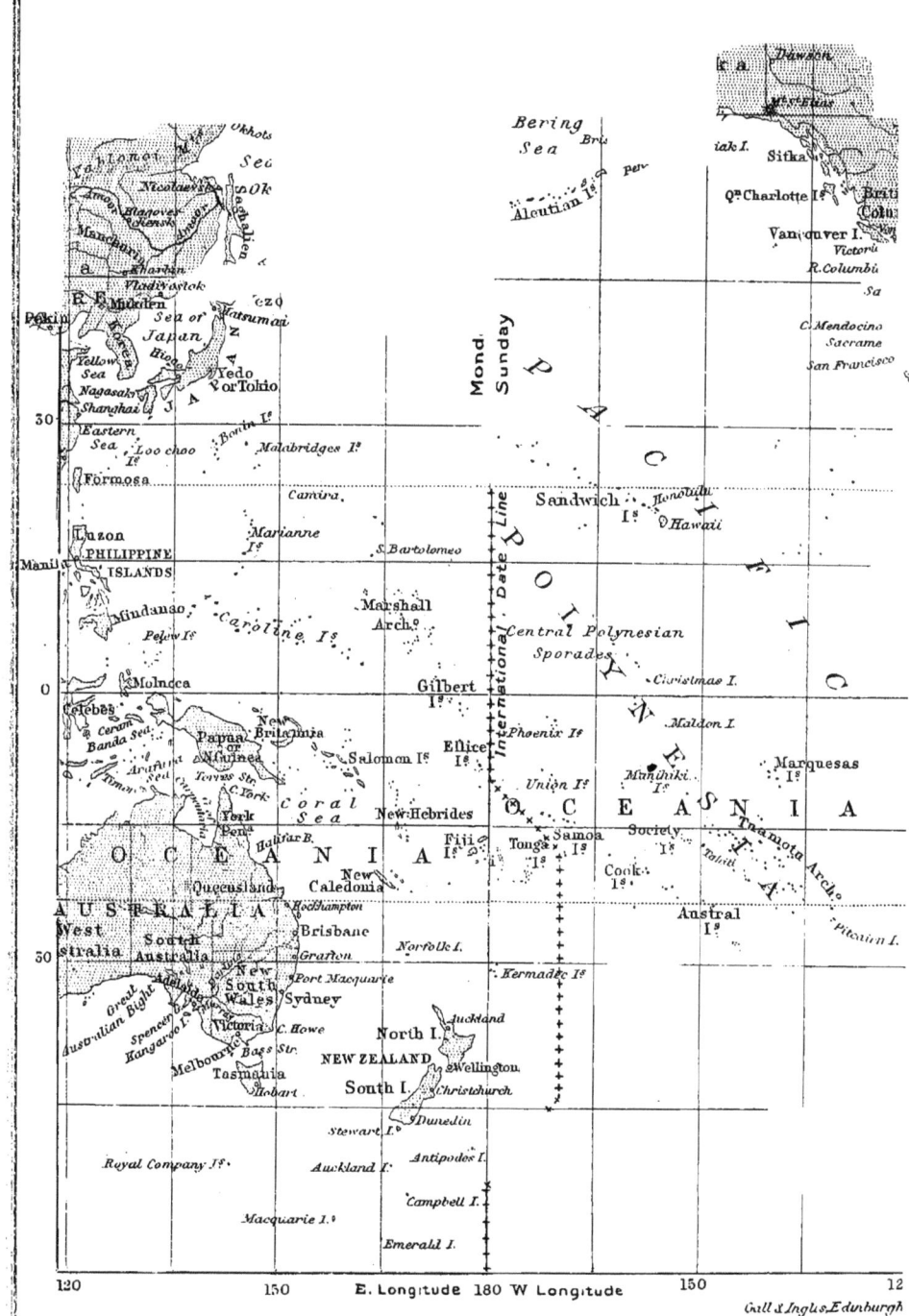

in respect that the Asiatic day is kept in these groups. Although these islands are numerous—the Friendly Islands alone being about 150 in number—they are comparatively unimportant. The day kept in these groups is, no doubt, decided by their neighbourhood to, and dependence on, New Zealand; which lies on the other side of the 180th meridian, in its near neighbourhood, and to which therefore the Asiatic day properly applies.

It will be seen that the divergences of the date line from the 180th meridian which still remain, are, in many cases, unavoidable, and that, so long as they are recognized and accounted for, they are of little practical importance. It may, notwithstanding these divergences, be considered as now virtually a matter of international understanding that the date changes, and the day begins at the 180th meridian, this meridian being the antipodes of that of Greenwich. Of course, in the extreme north and south, the adjustment of the line may still be regarded as not definitely settled, and as dependent, to some extent, on geographical investigation; but, in these parts, the exact position of the line is of small consequence.

In comparatively recent years a movement in favour of an international system of what is called "standard time" has arisen, and already it has attained much success. The idea seems to have originated in the United States, where the great range of the country in longitude makes it impossible to use the time of one meridian as applicable to the whole country. It arose from a desire to facilitate the international telegraph and railway traffic. The suggestion, which is based on the recognition of the meridian of Greenwich as the prime meridian for the whole Earth, is that the surface of the Globe should be considered as divided into zones; each zone having a width of 15 degrees of longitude. The mean solar time of the central meridian in each zone would apply to the whole of the zone. In fact, if we compare the world to an orange, the sections of the orange would typify the ideal time arrangement. Thus Greenwich time would be applicable to the district from 7½ degrees east to 7½ degrees west of the Greenwich meridian, a district including the whole of Britain and France and a great part of Ireland. The neighbouring zone, on the eastern side, would extend from 7½ degrees east to 22½ degrees east, and the time would be regulated by the meridian 15 degrees east, and would, therefore, be exactly one hour earlier than Greenwich time; while the corresponding zone to the west would be regulated by the meridian 15 degrees west, and would be exactly one hour later than Greenwich time. Ireland would thus be divided between two zones. In details like this the scheme is necessarily defective, but in its broad aspect it is very desirable.

In the United States four standard meridians were adopted in 1883, these being 75°, 90°, 105°, and 120° respectively, west of Greenwich, so that the clocks, in these respective zones, are either five, six, seven, or eight hours later than Greenwich mean time. These times are called Eastern time, Central time, Mountain time, and Pacific time, respectively. Later on a fifth zone was formed, in which the time—which is called Inter-Colonial time—is decided by the mean solar time at the meridian 60° west, and is accordingly four hours later than the mean time at Greenwich.

The adoption of standard time has now been legalized in all European Countries with the exception of Portugal. In Spain, where the system was adopted in the beginning of the twentieth century, the hours are counted from 0 hours to 24 hours, so that the use of the letters a.m. and p.m. is avoided. In Great Britian, France, Spain, Belgium, Holland, and Luxemburg, West European time, or the mean solar time of the meridian of Greenwich, is now in use. It Italy, Switzerland, Germany, Denmark, Norway and Sweden, Central European time, being the time of the meridian 15 degrees east of Greenwich, that is one hour earlier than Greenwich time, has been adopted. Russia, Roumania, Bulgaria, Turkey, and Egypt are subject to Eastern European time, which is two hours earlier than that of Greenwich—being regulated by the mean solar time of the meridian 30 degrees east of Greenwich, the meridian which also decides the time for British South Africa. In Australia, three time-zones have been arranged, which are regulated by the meridians 120°, 142½° (South Australia) and 150°, respectively, east of Greenwich, and which, therefore, fix the time as eight, nine-and-a-half, and ten hours respectively, in advance of Greenwich time. The last of these applies to Victoria, New South Wales, and Queensland. In Japan the time is now nine hours in advance of the mean time at Greenwich.

In other parts of the Globe where the advantages of standard time have been recognized, it has been found— as in South Australia—inconvenient to adopt a time differing from that of Greenwich only in the hour. The hour has therefore been divided by two, and it has been arranged to fix a local standard time differing from that

of Greenwich by thirty minutes, in addition to the required difference of the hour-hand of the clock. This is
the case in India and Ceylon, where the Standard time
adopted is five hours and thirty minutes ahead of Greenwich; in Burmah, where it is six-and-a-half-hours earlier
than it is in Greenwich; and in New Zealand, where the
time is eleven-and-a-half-hours before Greenwich time.
The system of five zones adopted in the United States
has also been made use of in Canada. Thus, the zone
system of standard time, referable to the meridian of
Greenwich. has, as far as practicable, been sanctioned in
almost all civilized countries.

In this connection it might be suggested that we ourselves might forward the system to a slight extent, in
accordance with the movement in other countries, by
fixing standard time for Ireland at thirty minutes later
than Greenwich time, instead of having a difference of
twenty-five minutes as at present. This would cause a
slight discrepancy in Irish time from the mean time of
the Dunsink Observatory, which is about four miles
north-west of Dublin. Irish time would then be that
applicable to the meridian $7\frac{1}{2}$ degrees west of Greenwich.
This is almost exactly the position in longitude of the
town of Tullamore, which lies about 48 miles to the
west of Dublin. In fact, as Dublin lies on the east
coast while the meridian 7° 30′ W. passes through
Ireland near the centre, a difference of thirty minutes
from Greenwich mean time would be more correctly the
time applicable to the whole country than that at
present kept.

The very general adoption of the meridian of Greenwich as the prime meridian of the Globe for the purpose

of computing standard time, and the international recognition of the meridian forming the antipodes to that of Greenwich as the place where, as far as possible, the change of day is effected, may be said to be a delicate and flattering international compliment to the British nation. By implication Great Britain is acknowledged as the power chiefly concerned in these world-wide questions.

The district through which the date line passes has certain advantages of a unique character which it can offer as attractions to the public. Even in this, the twentieth century, there are not a few who have their lucky and unlucky days, as Cromwell had his 3rd of September. There are many schoolboys also, and perhaps others, who have a decided preference for Saturday over Sunday. To all these the vicinity of the date line offers special inducements. If it is wished to get an unlucky day deleted from the calendar, all that is necessary is to take up a position immediately to the east, or American, side of the date line, on the preceding day, and, as midnight comes on and the day is gradually passing away, to cross the date line to the western, or Asiatic, side when, *mirabile dictu*, the dreaded day will be found to have had no existence, the succeeding day having already dawned. It will, of course, be necessary to abstain from recrossing the fateful line for at least twenty-four hours, as otherwise the hands of the clock will spring backwards, and the day which it was desired to avoid will show that it was only in concealment. If, on the other hand, it is desired to duplicate a lucky day, the reverse process has to be gone through. In the same manner, the schoolboy wishing to stretch out his weekly holiday may cross from the western, or Asiatic, side to the eastern, or American, side of the line, as Saturday is giving place to Sunday, and he will find that Saturday, instead of being finished, is only just beginning.   If he postpones his return for twenty-four hours he will find, on recrossing the line, that Monday has already begun.   To all who are anxious to escape the falling due of bills, or other obligations, the vicinity of the date line should prove a most fascinating resort, as they can so easily prevent the inauspicious date from ever, in their experience, having any existence.

We often claim for London, as the Capital of the British Empire, that it is situated at practically the centre of the land surface of the Globe.   The recognition of the 180th meridian as the international date line may be said to give London the position of being at the centre of what we may call the "earth day."   By this term we may truly describe the day which, for the time being, exists on the Earth.   It is different from the solar day and from the sidereal day.   It is the day which covers the Earth from the 180th meridian west-wards to the 180th meridian again.   Only for one instant, during the twenty four hours of the mean solar day, does the "earth day" coincide with the civil day. This is the moment of noon at London.   At that instant, if we can conceive of an imaginary line like the 180th meridian as having an appreciable breadth, we may say that the "earth day" is just entering at one margin of the line, and just passing away at the other.   At other times, the "earth day" will consist partially of two civil days.   At midnight in London the "earth day" consists, as regards the hemisphere to the east of

London, of the period from the noon which will not arrive in London for twelve hours to the midnight then present, and, as regards the hemisphere to the west of London, of the period from the midnight then present to the noon which passed over London twelve hours previously. In this "earth day," we may say that it is always "noon" in London, if by noon we understand the centre of the existing time. Therefore, in a map of the World, it would not be inappropriate to have the meridian of Greenwich as the central meridian, as this would be indicative of the important fact that, in the day which for the time being is passing over the Earth, from the 180th meridian in the east to the same meridian in the west, Greenwich, or London, is situated at an equal distance between its beginning on the one side, and its termination on the other.

THE ROTATION OF THE EARTH AND OTHER MEMBERS OF THE SOLAR SYSTEM.

SYNOPSIS.

The stupendous importance of the Earth's rotation and our ignorance as to its causes—Period of Earth's rotation and uncertainty as to its absolute constancy—Doubt as to rotational periods of *Venus* and *Mercury*—The rotational periods of other planets—Peculiar character of the Sun's rotation—The rotation of the Moon—Difference in rotational periods, although all subject to same laws—Future investigation—What known differences in the rotating bodies may have contributed to the difference in the rotational periods—Time of rotation invariably either very brief or equal to time of revolution—Consequences of rotational period of planets (as distinguished from satellites) coinciding with period of revolution—Moon's rotation different in character from Earth's rotation—Comparison of rotational periods of Sun and planets as evidenced by actual surface displacement—The same comparison taking each rotating body to be of same size as the Earth—The Nebular Hypothesis—The direction of rotation of Sun and planets—Any hypothesis as to causes of rotation at present purely speculative—A consistent hypothesis not necessarily correct—Causes effecting change of rotational periods — Contraction—Formation and thickening of crust—Internal heat and gradual cooling of interior—Sun in early stage of rotatory existence with rotational period lessening—Polar flattening as influenced by rotation—The period of the rotatory existence of the various planets—The peculiar character of the Moon's rotation—The three classes of rotational period—Possibility of existence of maximum and minimum axial velocity independent of period of revolution—Possibility of abrupt change from independent period of rotation to period coinciding with time of orbital revolution.

THE ROTATION OF THE EARTH
AND OTHER MEMBERS OF
THE SOLAR SYSTEM.

SIR ARCHIBALD GEIKIE, in the valuable treatise on geology which he has contributed to the *Encyclopædia Britannica* has the following sentence :—" In obedience to the impulse communicated to it at its original separation, the Earth rotates on its axis."

There is perhaps no phenomenon of Nature on which the interests of humanity are more absolutely dependent than the rotation of the Earth, and there are few phenomena as to the causes of which we are more profoundly ignorant. But for the rotation of the Earth on its axis there would be no recurrence of day and night such as we have under existing circumstances. In all probability, the Earth would be quite unsuitable for human existence. Yet all that we really know as to the source of this wonderful phenomenon is that it was communicated to the Earth at its origin, and that, in obedience to this original impulse, the Earth continues to rotate.

The period which the Earth takes to complete an axial rotation has been found to be 23 hours, 56 minutes, 4·098 seconds, that being the interval between successive returns of the same meridian to the same part of the heavens—the sidereal day.

No variation has been observed in this period since its first discovery. So far as human experience has shown, the time of the Earth's rotation is absolutely constant. This does not, however, indicate that it is not subject to change. The tendency of scientific opinion, indeed, is to conclude that the time occupied in the rotation has changed considerably during past ages, and that it will greatly change during the ages of the future. After all, the time during which a planet, such as the Earth, is fitted to be the abode of intelligent life is quite insignificant in comparison with the period of the planet's existence, so that, even if for centuries to come the period of rotation should continue to show no appreciable change, we could not conclude that it always has been, and will always continue, of uniform length.

It might be supposed that much information could be derived as to the laws governing the Earth's rotation from observation of, and comparison with, the corresponding movements of other planets. Most careful investigation has been made as to the rotational movements of the other planets, but the results arrived at are not such as can, with confidence, be employed in the investigation of the causes governing the Earth's rotation. The reason for this is that equally skilful observers, who have been equally favourably situated, and who have studied the matter with painstaking care, have, in different instances, arrived at most incompatible conclusions.

Venus is the most brilliant of all the planets, and in size is not much inferior to the Earth. Its orbit is immediately within that of the Earth, its period of revolution around the Sun being about 224 days, 17 hours. When nearest to the Earth, *Venus* is at a distance

THE LARGER PLA   S OF   H
Sh   re iv

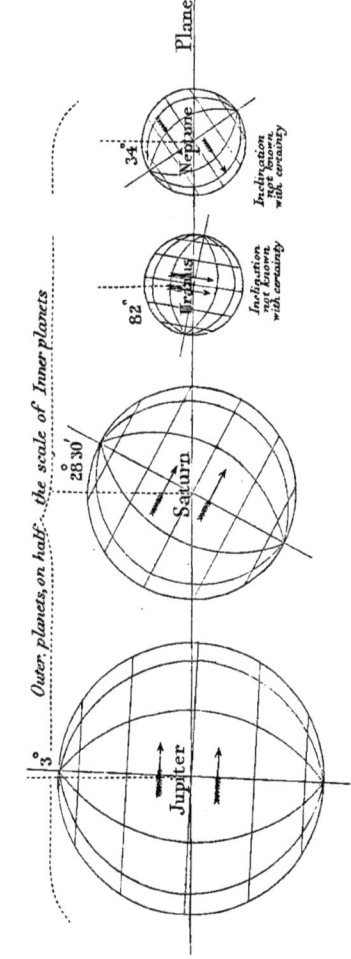

Outer planets, on half the scale of Inner planets

Jupiter 3°

Saturn 28 30'

Uranus 82°
Inclination not known with certainty

Neptune 34°
Inclination not known with certainty

Plane of planets orbit

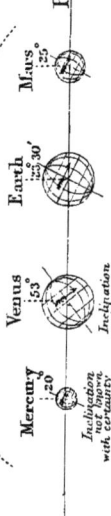

Inner planets, on twice the scale of the outer planets

Mercury 20
Inclination not known with certainty

Venus 53'
Inclination not known with certainty

Earth 23° 30'

Mars 25°

Plane of planets orbit

of about 26 millions of miles.   As might be expected, *Venus* has received much attention, and considerable information regarding it has been acquired.   The period of its rotation has formed a subject of special study, and what is the result?   Certain astronomers have arrived at the conclusion that the planet rotates on its axis in 23 hours, 21 minutes, 15 seconds, while others, just as eminent, have decided that the period of rotation is the same as the period of revolution, which, as mentioned above, is about 224 days, 17 hours.

Mercury is much smaller than *Venus*, and revolves around the Sun far within the orbit of *Venus*.   Observation of its movements is much more difficult.   Nevertheless, *Mercury* also has received very careful attention from many distinguished astronomers, and one subject of their special investigation has been the period of its rotation.   Near the beginning of last century, Schröter, an astronomer renowned, amongst other work, for his telescopic study of the planets, observed that one of the horns of *Mercury*, when its illuminated part appeared in the crescent form, showed an irregular marking at certain intervals.   Here was something to aid observation.   From repeated inspection, he came to the conclusion that the period of the planet's rotation was 24 hours, 4 minutes.

Harding, another skilful observer, arrived at almost exactly the same result, the time as found by him being 24 hours, 5 minutes, 28 seconds.

In recent years, however, Schiaparelli, of Milan, the great Italian astronomer, after lengthened study of the same subject, decided that *Mercury* rotates on its axis not in 24 hours, 5 minutes, 28 seconds, or in 24 hours,

4 minutes, but in exactly the same time as it revolves around the Sun, which is about 88 days.

Among the outer planets, *Uranus* and *Neptune* have been observed, but no definite conclusion has yet been arrived at as to the period of their axial rotation.

On the other hand, there seems to be general agreement as to the time, or approximate time, taken by the remaining larger planets of the solar system in their axial rotation. These planets are *Mars, Jupiter,* and *Saturn,* and the respective periods are as follows :—

Mars,	...	24 hours, 37 minutes, 23 seconds.
Jupiter,	...	9 hours, 55 minutes, 26 seconds.
Saturn,	...	10 hours, 14 minutes, 24 seconds.

In the case of *Saturn,* indeed, there is some difference of opinion amongst astronomers as to the exact time of rotation, but the discrepancy is slight, and goes to show with what extreme care these subjects are investigated. It is said that Herschel, in 1794, found the period to be 10 hours, 16 minutes. Proctor, in the *Encyclopædia Britannica,* gives it as 10 hours, 29 minutes, 17 soconds. Professor Asaph Hall, the great American astronomer, concluded, from personal observations, that it was 10 hours, 14 minutes, 24 seconds. M. Camille Flammarion, a celebrated French astronomer, found it to be 10 hours, 14 minutes, 14 seconds, and Mr W. F. Denning, who is an indefatigable observer of the heavens, obtained, in 1903, a period of 10 hours, 37 minutes, 56·4 seconds. Although, therefore, the time we have accepted as correct, which is the time obtained by Professor Asaph Hall— 10 hours, 14 minutes, 24 seconds—may not be absolutely accurate, it is evidently very nearly so, and it is recognized by several recent authorities.

Each of these three planets—*Mars, Jupiter* and *Saturn*—revolves in an orbit outside that of the Earth, the orbits increasing in the order given.

It is known that the Sun also, like the planets, rotates on its axis, and much information on this subject has been obtained by the study of the motion of Sunspots. It has been definitely proved that the rotation of the Sun differs in an important respect from the rotation of

TABLE SHOWING THE ROTATIONAL PERIOD OF THE SUN IN DIFFERENT LATITUDES.

Solar Latitude.	Period of Rotation.		
50° North.	27 days,	10 hours,	41 minutes.
30° ,,	26 ,,	9 ,,	46 ,,
20° ,,	25 ,,	17 ,,	8 ,,
15° ,,	25 ,,	9 ,,	10 ,,
10° ,,	25 ,	3	29
5° ,,	25 ,,	0 ,,	42 ,,
0° (Equator)	24 ,,	2	11 ,,
5° South.	24 · ,,	23 ,,	18 ,,
10° ,,	25 ,,	5 ,,	35 ,,
15° ,,	25 ,,	13 ,,	31 ,,
20° ,,	25 ,,	17 ,,	52 ,,
30° ,,	26 ,,	12 ,,	50 ,,
45° ,,	28 ,,	11 ,,	0 ,,

the planets, in that the rotational period varies in different solar latitudes. The Sun does not rotate as a whole—a fact which seems to prove that the Sun (at least so far as visible to us) is not a solid body, but is in a fluid state. The rotation is most rapid at and near the solar equator, where it is completed in 24 days, 2 hours, 11 minutes. The speed of rotation noticeably diminishes as the distance from the equator is increased, but it is noteworthy that it is not the same for corresponding latitudes. The Table on page 273, showing the rotational period for different solar latitudes, is given by Mr. R. A. Proctor, the well-known astronomical authority.

The Moon, too, has been found to rotate on its axis, and there is complete agreement as to the period of its rotation. Indeed, one writer on astronomy expresses the opinion that the time of the lunar rotation is the most remarkable circumstance known about the Moon. The period is exactly the same as that taken by the Moon to accomplish its revolution around the Earth, which is 27 days, 7 hours, 43 minutes, 11·545 seconds. From this it follows that the Moon always presents the same side towards the Earth, subject to the changes occasioned by the variation in its height in the heavens, and in its eastward or westward direction—its librations in latitude and longitude.

Thus, in the solar system, even in respect of the small number of bodies of which the rotational periods are known with some degree of certainty, there is very considerable divergence in the periods.

The time required by the Moon to accomplish the movement is very nearly as long as that required by the Sun in its higher latitudes, although the size of the Moon is quite insignificant in comparison with the size of the Sun. As regards the Earth, *Mars, Jupiter,* and *Saturn,* on the other hand, the periods of rotation, although they vary considerably if compared with each other, are wonderfully similar as regards the actual time occupied. The time difference between *Mars* with the longest period, and *Jupiter* with the shortest, is only 14 hours, 41 minutes, 57 seconds.

In a matter like this, where no such uncertainty exists, it may be rash to make any assumption, but yet there does not seem to be any reason to doubt that the same laws apply to the rotation of all the heavenly bodies, not only in the solar system, but throughout the sidereal universe.

Doubtless, in the future, through patient investigation and the accumulation of observed facts, these laws will be discovered, and our descendants will then be in a position to judge as to the changes which time will bring about in the period of the Earth's rotation, and to know whether the length of the day is now different from what it was in the times of our ancestors. Although this discovery may yet be distant, it is not uninteresting to consider the information already obtained in regard to the rotation of the various members of the solar system, and to see how the period of rotation of one body compares with the period of another, and what known differences between these bodies *may* have contributed towards the existing difference in the rotational periods.

In the observations of *Mercury* and *Venus*, to which reference has been made, the extreme difference in the results arrived at by equally eminent men as to the periods of the rotation of these planets is a most noteworthy fact. The discrepancy is accounted for by the indistinctness of the markings on the disks of these planets. It is, however, extraordinary that in both cases the conclusions reached would either place the planets in the same class as the Earth, *Mars*, *Jupiter* and *Saturn*, with a very short period of rotation, or in the same class as the Moon, with a period corresponding exactly to the time of revolution. It is remarkable that, so far as either certain or uncertain conclusions have been arrived at, the actual time of rotation is invariably either very brief, or is of the same length as the period of revolution.

In so far as the information as to planetary rotation is *definitely* ascertained, the Moon is in a class by itself. In every lunar revolution there is an exact lunar rotation. Now the Earth is in precisely the same relation to the Moon as the Sun is to the Earth. The Sun is the primary which controls the Earth's revolution; the Earth is the primary which controls the Moon's revolution. The Moon certainly revolves around the Sun just as the Earth revolves around the Sun, but there is a difference in the nature of their solar revolutions. The Earth revolves around the Sun as the main source of its attraction, the Moon does so in following the Earth as the main source of its attraction. If, as is generally supposed to be the case, the whole Solar System is revolving around an unknown centre, then this unknown centre has the same relation to the Earth as the Sun has to the Moon. The Earth revolves around such a centre simply through following the Sun as the primary which controls its orbital revolution.

Let us suppose that the Earth, in revolving around the Sun, had the same kind of axial rotation as the Moon has in revolving around the Earth, and see what the result would be. In this case the Earth would take about 365¼ days to accomplish the rotation. It is obvious that the alternation of day and night would, if the period of rotation were the same as the period of revolution, cease to exist. One side of the Earth would be always turned towards the Sun and have perpetual day; the other side would be always turned away from the Sun and have perpetual night. A slight modification would no doubt result from the obliquity of the ecliptic, whereby the regions within the Arctic and Antarctic circles would, to a certain extent, have a half-yearly alternation of light and darkness, and this, in a less degree, would also apply to the neighbouring parts of the Earth, but, subject to this qualification, the opposite hemispheres of the Earth would evidently have either continuous day or continuous night.

Of course, such a state of matters does not occur in our Satellite. There is, in the Moon, a regular recurrence of light and darkness, each in turn lasting for about $14\frac{1}{2}$ of of our solar days. This, however, is simply because the primary around which the Moon revolves is not the source of its light and heat. Its revolution around the Earth happens to bring it into such a position as to expose it to the light and heat of another body, with which its orbital revolution is not immediately concerned. A similar result could not follow in the case of the Earth, were the Earth's rotation similar to that of the Moon, simply because the centre of the Earth's revolution is also the source of the Earth's light and heat, the source, in fact of its day as distinguished from its night. If then the time required by the Earth to rotate on its axis were exactly the same as the time of its orbital revolution, it is clear that the axial rotation *as the cause of the alternation of day and night* would have ceased to exist. In this respect an axial rotation which coincides with the period of orbital revolution is of a totally different character from an axial rotation with an independent period.

The Moon's rotation, therefore, differs in character from the Earth's rotation. It might, indeed, be said that, in a certain sense, the axis of the Moon's rotation is not a line passing from pole to pole through the centre of the Moon, as the terrestrial axis of rotation passes from Pole to Pole through the centre of the Earth, but that it is a line passing through the centre of the lunar orbit.

Thus, as we have mentioned, *so far as observation has revealed*, every planet of the solar system rotates in either a very brief period, or in a period coinciding with that of its orbital revolution. Either the planet has the regular recurrence of day and night in an interval of time not dissimilar from that which applies on the Earth, or it has (as regards almost the whole of the surface of the planet) continuous day or continuous night.

The Sun indeed, appears, as regards its rotation, to fall into a different category from the planets. This is certainly the case as regards the different rates of rotation in the different latitudes. Let us see whether it is also the case in regard to the velocity of rotation.

To enable us to judge of the velocity with which these different bodies rotate, it is necessary to have the equatorial circumference in each case. The speed of rotation is indicated by the displacement occasioned thereby of any position on the equator during a given time.

The circumference of the Earth at the equator is about 25,000 miles, and the time of the Earth's rotation is 23 hours, 56 minutes, 4 seconds. Therefore every spot on the terrestrial equator is, by the axial rotation, caused to pass through about 25,000 miles in the time mentioned.

This gives an hourly displacement of about 17·4 miles a minute.

The diameter of *Saturn* is given by Mr Thomas Heath, of the Royal Observatory, Edinburgh, in a recent work· as 75,600 miles, which is somewhat greater than the length given by earlier writers. As the diameter of a circle is to the circumference as 1 is to 3·14159, this would make the length of the circumference of *Saturn* about 237,504 miles. Assuming that the time of the planet's rotation is 10 hours, 14 minutes, 24 seconds, this would make the displacement at the equator, caused by the rotation, 6·44 miles every second, which would be about 386·5 miles a minute.

The equatorial diameter of *Jupiter* is said to be about 87,700 miles, which would make the circumference about 275,517 miles. As we have seen, the time of rotation is 9 hours, 55 minutes, 26 seconds, so that the equatorial displacement produced by the rotation is about 7·71 miles a second, or 462·7 miles a minute.

Mars, the nearest planet moving in an orbit outside the Earth, is about 4,180 miles in equatorial diameter, and has therefore a circumference of about 13,131 miles at the equator. The rotational period is 24 hours, 37 minutes, 23 seconds, which makes the distance which each point on the Martial equator passes through in a minute, by the axial rotation of the planet, about 8·9 miles.

If we take the time of rotation of *Venus* and *Mercury* as being 23 hours, 21 minutes, 15 seconds; and 24 hours, 5 minutes, respectively, we can similarly obtain the surface displacement at the equators occasioned by their rotation. The diameter of *Venus* is about 7,700 miles,

which makes the circumference about 24,190 miles, so
that the displacement at the equator caused by the
rotational movement will be nearly 17·3 miles a minute.
The diameter of *Mercury* is about 2,960 miles, making the circumference about 9,299 miles, so that the
corresponding displacement is about 6·4 miles a
minute.

Let us now make a similar calculation in regard to the
Sun, confining our attention to the period of rotation at
the equator, in view of the period differing materially at
different latitudes. At the solar equator the time of
rotation is 24 days, 2 hours, 11 minutes. The Sun's
diameter is estimated to be about 866,500 miles. This
makes the length of the circumference about 2,722,190
miles, and the surface displacement at the equator, arising
from the rotation, about 78·4 miles a minute.

The following Table shows the distance which is
travelled by a point on the equator of the Sun, and each
of these planets, in the space of one minute, through the
effect of the axial rotation, taking the time of the
rotation of *Mercury* and *Venus* to be respectively 24
hours, 5 minutes, and 23 hours, 21 minutes, 15 seconds.

Jupiter,	...	about 462·7 miles a minute.	
Saturn,	...	„ 386·5 „	„
Sun,		„ 78·4 „	„
Earth,	...	„ 17·4 „	„
Venus,	...	„ 17·3 „	„
Mars,	...	„ 8·9 „	„
Mercury,	...	„ 6·4 „	„

It is noticeable that the range of variation is excessive,
and it is also apparent that, if the Sun is disregarded,
the axial velocity is exactly according to size. Thus, by

regular gradation, the velocity of rotation falls from *Jupiter*, the giant planet, to *Mercury*, the smallest of the group. It would seem that all the circumstances indicate specialities in the case of the Sun. It does not rotate as a whole; it is not in the solid state on the surface; and it does not fall into the natural order of size observed by the planets in their rotational velocity. We may therefore, in the meantime, leave the Sun out of account, and give our attention exclusively to the planets,

Although the velocity of rotation is according to size, this has, to a certain extent, to be looked for. It is evident that as the circumference increases so does the surface velocity of rotation, even though the actual time occupied in the rotation is unchanged. If the Earth were swollen to the size of *Jupiter*, while the time of rotation continued as at present, each point on the surface at the equator would be carried by the rotatory movement through a much greater distance in a minute of time than it at present is; and, in the same way, if *Jupiter* were contracted to the size of the Earth, its time of rotation continuing unchanged, the displacement at the equator occasioned by the rotation in a given time would be very greatly lessened.

Let us now see what distance a point on the equator of each of these planets would be displaced by the axial velocity in a minute of time, supposing each planet to be the same size as the Earth, and the respective times of rotation as at present—assuming that *Mercury* and *Venus* rotate in the respective periods on which our former calculations are based. The times are as given on the following page.

Planet.				Axial velocity at equator, if planet of same circumference as the Earth.
Jupiter,	...	...	...	41·98 miles a minute.
Saturn,	...	...	...	40·69 ,, ,,
Earth,	...	...	...	17·41 ,, ,,
Venus,	...	...	...	17·84 ,, ,,
Mars,	...	...	...	16·92 ,, ,,
Mercury,	...	...	...	17·30 ,, ,,

It is interesting to see how these planets divide themselves into two groups, consisting of the outer and inner planets respectively. The speed of rotation of the individual planets forming each group is wonderfully similar. Thus the inner planets from *Mercury* to *Mars* have an average velocity of 17·37 miles a minute, and the extreme variations from this mean are only a fraction of a mile per minute—·47 of a mile of greater velocity in the case of *Venus*, and ·45 of a mile of less velocity in the case of *Mars*. The two outer planets have a mean velocity of 41·335 miles a minute, the individual difference from the mean being ·645 of a mile a minute, and, as we have only two planets to work on, the difference is, of course, the same for each.

The Nebular Hypothesis has some bearing on our present subject. According to Sir Robert Ball:—

"Laplace supposed that our Sun had once a stupendous nebulous atmosphere which extended so far out as to fill all the space at present occupied by the planets. This gigantic nebulous mass, of which the Sun was only the central and somewhat more condensed portion, is supposed to have a movement of rotation on its axis. There is no difficulty in conceiving how a nebula, quite independently of any internal motion of its parts shall also have had as a whole a movement of rotation. In fact a little consideration will show from the law of probabilities that it is infinitely probable that such an object should really have *some* movement of rotation, no matter by what causes the nebula may have originated. As this vast mass cooled, it must by the laws of heat have contracted towards the centre, and as it contracted it must, according to a well-known law of dynamics, rotate more rapidly. The time would then come when the centrifugal force on the outer parts of the mass would more than counterbalance the attraction of the centre, and thus we would have the outer parts left as a ring. The inner portion will still continue to contract, the same process will be repeated, and thus a second ring will be formed. We have thus grounds for believing that the original nebula will separate into a series of rings all revolving in the same direction with a central nebulous mass in the interior. The materials of each ring would continue to cool and to contract until they passed from the gaseous to the liquid condition. If the consolidation took place with comparative uniformity, we might then anticipate the formation of a vast multitude of small planets such as those we actually do find in the region between the orbit of *Mars* and that of *Jupiter*. More usually, however, the ring might be expected not to be uniform, and therefore to condense in some parts more rapidly than in others. The effect of such contraction would be to draw into a single mass the materials of the ring, and thus we would have a planet formed, while the satellites of that planet would be developed from the still nascent planet in the same way as the planet itself originated from the Sun. In this way we account most simply for the uniformity in the direction in which the planets revolve, and for the mutual proximity of the planes in which their orbits are contained. The rotation of the planets on their axes is also explained, for at the time of the first formation of the planet it must have participated in the rotation of the whole nebula, and by the subsequent contraction of the planet the speed, with which the rotation was performed, must have been accelerated."

The *direction* of rotation of the Sun and of each of the planets mentioned appears to be similar. The Earth rotates eastwards so that if an observer in *Mars*—or at least in the Northern Hemisphere of *Mars*—were watching the Earth's rotation he would see any surface-marking upon which he had fixed his gaze come into view on his left, pass across the Earth's disk, and disappear from view on his right. This is what happens when Sunspots are observed, so that the Sun also appears to rotate in the same direction.

In each of the planets mentioned (at least in and near their equatorial regions), the Sun will appear to rise in an easterly direction and to set in a westerly direction.

It is thought, however, that, in the case of the two more distant planets, *Uranus* and *Neptune*, whose periods of rotation are quite unknown, this does not apply. The direction of the rotatory movement of these two planets seems to differ considerably from that of the Earth and the other planets. The rotation of *Neptune* is supposed to be practically in the reverse direction from the rotation of the Earth so that the apparent motion of the Sun will also be reversed. The Sun will appear in *Neptune* to rise towards the west and to set towards the east. In *Uranus* again the poles of the planet's axis are believed to be not far removed from the plane of its orbit, a position which in the case of the Earth, is more nearly occupied by the plane of the equator. The Sun will, in *Uranus*, appear to rise in what to us would seem to be a northerly direction, and to set in a southerly direction. The information respecting *Uranus* and *Neptune* is, however, by no means certain.

In the present state of information as to the rotation

of the various members of the solar system, any theory as to the causes of the rotatory movements, or as to the laws which govern these movements must necessarily be purely speculative. The utmost that can be definitely stated is that the known facts are either consistent or inconsistent with certain hypotheses.

Sir Robert Ball gives us certain suggestions as to the possible origin of the planetary rotations, in the passage as to the Nebular Hypothesis which we have quoted. He however goes on to say :—

"Should any one be sceptical as to the sufficiency of these laws to account for the present state of things, science can furnish no evidence strong enough to overthrow his doubts until the Sun shall be found growing smaller by actual measurement, or the nebula be actually seen to condense into stars and systems."

If, now, any supposition were put forward, as to the rotatory movement, inconsistent with a *single fact* of undoubted truth, this fact would in itself prove the erroneousness of the supposition, but, on the other hand, should any supposition be consistent with the known facts, its truth cannot, in the present state of our knowledge, be thereby assumed. If it were suggested that the rotatory movement of the planets is occasioned simply by their rolling along in their orbits as the wheel of a railway carriage rolls along the track, the fallacy of the suggestion is at once evident from its being inconsistent with the motion of the Earth itself. If this suggestion were feasible, then each planet would during its rotation advance in its orbit the length of its own circumference. The Earth consequently would in the time of rotation advance about 25,000 miles. The Earth, however, actually moves forward in its orbit about 18½ miles every second, or more than 1½ *millions* of miles in the time it completes a rotation.   Thus, although it happens that in the case of *Saturn* the displacement in the equatorial regions occasioned by the rotatory movement has some similarity to the amount of orbital progression in a corresponding time, we do not require to go outside the information gained as to the Earth's own movements to be satisfied that the suggestion is untenable.

In the same manner it would appear that the rotatory movement has no direct dependence on the orbital movement.   If it had, then we should expect the speed of rotation to vary with the changes in the orbital velocity, and this is not the case.

There does not, however, at present, appear to be any inconsistence in the supposition that the rotatory movement may have some relation to, and some dependence on, the condition of each individual planet as regards heat.   It would, indeed, seem to be quite possible that the speed of rotation is mainly decided by the state of each planet in the matter of self-centred heat.

We have seen that, according to the nebular theory, the rotatory movement of planets had its origin in the nebula before the planets themselves had any separate existence.   It is believed that the planets, when they came into existence, retained the rotation which characterized the nebula out of which by condensation they were gradually formed.   By the subsequent, and very gradual, contraction of each planet, this movement of rotation would, according to the laws of dynamics, become by degrees more rapid, the process of acceleration going on, very probably, through milleniums.   In the course of ages, each planet's outer surface would become solid, the solidity slowly extending inwards and further contraction being rendered difficult, if not absolutely prevented. The speed of rotation would reach its maximum on the cessation of contraction of the planet as a whole.

With the solidification of the surface, and the ceasing of further contraction of the planet in its entirety, another set of circumstances may come into play. The rotatory movement at this stage may be supposed to be continued mainly through the influence of the heated interior. Imperceptibly, but surely, the crust must thereafter thicken, and the interior itself become colder. The rotational period may, as the crust thickens, gradually increase in length. The duration of the day and the duration of the night may, by slow degrees, become greater, until, at last, the regular recurrence of day and night shall cease, and almost a hemisphere will have continuous day, while an equal area will have continuous night. It would seem not improbable that a change from the recurrence of day and night to an era of continuous day and continuous night may take place with some abruptness. If, as we have conjectured, the main operating cause in the former stage of planetary rotation is the self-centred heat of the planet, it would seem that a time must come, when, through the gradual loss of heat, what we may describe as "the internal mechanism" shall be unequal to the work required of it, and there shall occur a more or less sudden breakdown. When this occurs, the main operating influence in relation to the rotation may be supposed to pass from the planet to its primary—the centre of its orbital revolution, which

in the case of the Earth and the other important planets
of our system, is the Sun.   The planet would, no doubt,
still continue to rotate on its axis, but the time of
rotation would thenceforward exactly correspond with
the time of the orbital revolution.

In this connection the Sun may be considered as being
in a comparatively early stage of rotatory existence—a
stage at which no crust has yet been formed on the
surface.   According to general astronomical opinion the
Sun, is composed of a central mass of intensely heated
matter surrounded by a shell of luminous clouds.   From
centre to surface the Sun is supposed to be in a gaseous
state, although, no doubt, the gaseous vapour of which it
is constituted must be under enormous pressure.   From
the intensity of the internal heat, it is believed that even
the centre of the Sun, although the densest part of the
solar globe, must necessarily, notwithstanding the pressure
to which it is subjected, be in the gaseous form.   The
Sun, then, is evidently in a stage at which contraction
will take place most rapidly, although the word
"rapidly" applies only in comparison with other
heavenly bodies.   It is, indeed, supposed that it is
through the gradual contraction of the Sun that the
enormous quantities of heat which are constantly
radiating from the solar globe are generated.   The Sun
is supposed to be decreasing in diameter by somewhat
less than one hundred yards annually, or about four
miles every century.   Although the diameter of the Sun
is about 866,500 miles in length, this constant decrease
will necessarily, in the lapse of ages, cause an appreciable
diminution in the size of the great centre of our system.
As the contraction of a heavenly body must, according to

the passage which we have quoted from Sir Robert Ball, induce more rapid rotation, the rotatory speed of the Sun must at present be on the increase.

We have seen that the velocity of rotation of the Sun is now, as indicated by the displacement occasioned thereby of any point at the solar equator in a given time, very much less than is the rotatory velocity of *Jupiter* or *Saturn*. From the fact that these large planets complete a rotation in less time than any of the smaller planets, it would appear that increased size is no obstacle to brevity of time of rotation. Thus, the fact that the Sun is so much larger than any planet would seem to be no proof that its axial rotation must necessarily occupy a longer time. In fact, the evidence, if of any weight, is rather in the opposite direction. The actual time of rotation of the four planets whose period is known with certainty, becomes less as the size of the planet increases. *Mars*, the smallest of these planets, takes the longest time to rotate, and *Jupiter*, the largest, rotates in the shortest time. The Earth is larger than *Mars*, and its time of rotation is shorter. *Saturn* is smaller than *Jupiter*, and its time of rotation is longer. There is, therefore, no recognizable difficulty in the period of the Sun's rotation being very much less than it actually is; and all the evidence available appears to show that the Sun is now at the stage of existence when the time of rotation is gradually lessening.

In connection with the Sun, it is a most interesting peculiarity that the time of rotation varies with the latitude, being shortest at the equator, and longest at the highest latitudes. There seems to be no reason against the supposition that this peculiarity may have occurred also in the case of the various planets in long past ages, when they were in the primal stage of planetary existence. If, when the planets were formed, they were composed of gaseous matter, as the Sun still is, and if this matter had—as there seems to be no reason to doubt was the case—a movement of rotation, then, if the period of rotation were uniform for the whole mass forming the planet, it would seem that there would be a tendency for the planet to flatten out; so that, while circular at the circumference, the rotating mass would be absolutely flattened at the top and bottom. Any such tendency towards flattening would be greatly reduced if the rotating mass which formed the planet had, like the Sun, different velocities of rotation in different latitudes. The Earth and other planets are known to be somewhat flattened at the poles, but not to any great extent, so that it would seem not unlikely that the excessive flattening which rapid rotation would induce when the mass was in the gaseous or liquid state, was, to some extent, obviated by variation in the rotational velocity.

Of course, on the other hand. if the rotatory movement were not very rapid the tendency towards extensive polar flattening would not be so great. It seems not improbable that the different rotational speeds which exist in different latitudes in the Sun have an effect in preserving the sphericity of the solar orb, the figure of the Sun being, it is believed, perfectly spherical.

All the planets appear to differ from the Sun as regards the present stage of their rotatory movements.

We have noticed that the speed of rotation of the giant planets, *Jupiter* and *Saturn*, greatly exceeds that of the smaller planets. We may disregard the excessive equatorial-displacement occasioned by their huge size, and consider the velocity as if they were the same size as the Earth. Each part at the equator on the Earth rotates at the speed of about 17·4 miles a minute, while, if *Jupiter* and *Saturn* were of the same size as the Earth, the corresponding rotatory displacement would be 41·98 miles, and 40·69 miles respectively. On this basis the difference in the velocities is not extreme. Even in *Jupiter*, where the rotational velocity is greatest, it is not equal to $2\frac{1}{2}$ times the velocity of the Earth, and in *Saturn* the velocity is apparently slightly more than $2\frac{1}{4}$ times that of the Earth. If *Jupiter* and *Saturn* were of the same size as the Earth, the equatorial displacement occasioned by the rotation would in these planets be roughly about $2\frac{13}{32}$ miles, and $2\frac{11}{32}$ miles respectively, in the same time as it would on the Earth be about one mile.

Now, *Jupiter* and *Saturn* would each appear to be but little removed from the gaseous or the liquid state. Although a crust has formed on the surface of each of them, it would appear that the thickness of the crust is still very slight. In the case of *Saturn*, as the smaller planet, cooling will take place somewhat more rapidly than in *Jupiter*, and probably the surface crust is there of greater thickness than is the case in *Jupiter*.

There does not seem to be any inherent inconsistency in the supposition that *Jupiter* may at present be not far distant from the stage at which the velocity of rotation is most rapid. We have seen that, with planetary contraction, the speed of rotation increases although the mass is still in a gaseous state. We have suggested that when, through the radiation of heat, a crust becomes

formed on the surface, other circumstances probably
come into play, and that the speed of rotation may be
be lessened.

On these suppositions it would seem not unlikely that
the rotation of *Jupiter* is at present not much removed
from the most rapid which occurs in planetary existence.
There seems nothing improbable in the conjecture that
the speed of rotation must have a limit which is never
exceeded by a planet whatever the size of the planet
may be.   In fact, as we have seen, the actual speed of
rotation does not appear to be, to any noticeable extent,
affected by the size of the planet.

If, then, *Jupiter* rotates at nearly the utmost speed
which is attained in a planet's existence, *Saturn* may
have made some progress in the downward direction.
The equatorial diameter of *Jupiter* is about 87,700 miles,
while that of *Saturn* is about 75,600 miles.   *Saturn*
will therefore, if formed about the same time as *Jupiter*,
be now somewhat more advanced as regards crust formation on the surface.   This would harmonize with the
facts of the rotation periods.   While *Jupiter* rotates in
9 hours, 55 minutes, 26 seconds, with an equatorial
displacement (if its size be reduced to that of the Earth),
of about 41·98 miles a minute, *Saturn* requires 10 hours,
14 minutes, 24 seconds, and the corresponding displacement is about 40·69 miles.   As both *Jupiter* and *Saturn*
revolve at an enormous distance from the Sun, no
appreciable difference could arise between them in regard
to the prevention of the loss of internal heat by the
impact of solar heat.

Let us now glance at the inner planets.   Their times
of rotation and the equatorial displacement occasioned

by the rotation (supposing each planet to be of the same size as the Earth), may be taken to be as follows :—

Planet.	Time of Rotation.	Equatorial displacement if planet equal to the Earth in circumference at equator.
Mars,	24 hours, 37 minutes, 23 seconds,	16·92 miles a minute.
Earth	23 „ 56 „ 4 „	17·41 „ „
Venus,	23 „ 21 „ 15 „	17·84 „ „
Mercury,	24 „ 5 „ 0 „	17·30 „ „

Owing to the doubt as to the periods of the rotation of *Venus* and *Mercury* we can deal confidently only with the periods of the Earth and *Mars*. In their case our conjecture as to the period of rotation being influenced by the condition of the planet as regards internal heat, appears to agree with the actual facts. *Mars* is smaller than the Earth, and its orbital revolution occurs at a greater distance from the Sun. The equatorial diameter of *Mars* is about 4,180 miles in length, while that of the Earth is about 7,926 miles. If these two planets had their origin about the same time, *Mars* must be considerably more advanced in the cooling process than the Earth, and the crust will consequently extend a greater distance below the surface of the planet. The rotational speed of *Mars* would therefore, according to our conjecture, be appreciably slower than that of the Earth, and this is actually the case.

Venus is very similar to the Earth in size, its equatorial diameter being about 7,700 miles in length. Its orbital course is also within that of the Earth. If proximity to the Sun, with the consequent reception of solar heat, has any effect in lessening the rate of cooling, we should expect two diverse causes to affect the rotation of *Venus*. As it is somewhat smaller than the Earth, it will cool

somewhat quicker, and its crust—if it came into existence about the same time as the Earth—will be somewhat thicker, but the cooling process and the thickening
of the crust may be retarded by its orbital position. If
the time given in the foregoing table is correct it would
appear that *Venus* rotates slightly more rapidly than
the Earth, which on these suppositions could be
reasonably accounted for in conformity with our
conjecture.

Mercury is the smallest of the noticeable planets, its
diameter—according to Mr Thomas Heath—being 2,960
miles in length. Its mean distance from the Sun, in its
orbital course, is about 35,392,000 miles, while the
Earth's mean distance is about 91,430,000 miles. Two
diverse causes may therefore possibly affect the state of
this planet also as regards internal heat.   From its size
it may be judged that, if it had its origin at about the
same time as the Earth, it must be very greatly advanced
in the process of solidification, and thus, according to our
supposition, in the process of retardation of orbital
velocity.   What weight, if any, should be given to the
diverse cause—its proximity to the Sun—it is impossible
to judge. According to our table, the relative equatorial
displacement occasioned by the rotational movement is
—if the period of rotation is accepted as 24 hours, 5
minutes—slightly less in the case of *Mercury* in a given
time than in the case of the Earth.   The time of
rotation of *Mercury* is however exceedingly uncertain,
and if it were permissible to argue from our conjectural
argument itself, there would appear *ex facie* to be much
more likelihood that the rotational period of *Mercury*
(unless that planet came into existence much later than

the Earth), is identical with the period of revolution, than that this is the case with *Venus.*

We have noticed that the Moon's axial rotation is of a peculiar character; that if the planets, in their revolutions around the Sun, rotated on their axes in the same way as the Moon does in its revolution around the Earth, there would, in the planets, be an almost total absence of the alternation of day and night, this being a necessary consequence of the time of rotation being the same as the time of revolution, when the primary is also the source of light and heat.

In these members of the solar system with which we have dealt there thus appears to be some reason to suppose that the rotational periods fall into three different classes. We have, first, the Sun, in which the probability seems to be that the period of rotation will become shortened, and that ages may yet elapse before the minimum is reached and one period applies to the whole orb. We have, next, the planets *Jupiter, Saturn,* the Earth and *Mars* (and possibly *Venus* and *Mercury*) in which the minimum period of rotation has been already reached, and in which, during future ages, the probability is that the time of rotation will be increased. Of these *Jupiter* may at present be at, or only shortly beyond, the minimum period, while the other planets have in varying degrees progressed in the slowing-down stage. In the third class we have the Moon (and possibly *Venus* and *Mercury*) in which axial rotation with an independent period has come to an end, and the power associated with such internal heat, if any, as still remains, is insufficient to cause the planet's rotation.

If there exists a maximum limit to the axial velocity of the planets, as seems not unlikely, it is also possible that there is a minimum limit to the velocity required to produce an axial rotation having a period independent of that of the orbital revolution. We can scarcely suppose that if the axial velocity of, for instance, the Earth is decreasing, it will continue through future milleniums to decrease continuously at an infinitesimal rate, until at length, the period of the axial rotation coincides with the period of the orbital revolution. We have seen that there is some reason to believe that the maximum axial velocity of planets does not very greatly transcend the axial velocity of the Earth, It may be that the minimum velocity required to produce a rotation the period of which does not coincide with the period of orbital revolution, is not very greatly less than the Earth's speed of rotation. So far as investigation has revealed, the range of axial velocity, when reduced to one basis as regards planetary circumference, is not very great. Quite possibly any self-centred rotatory impulse, if of sufficient strength to cause the planet's rotation, must effect the rotation within a certain brief period. We can quite conceive that otherwise the retarding influences, such as tidal changes, would gain the mastery. Thus it may be, that, at the last, the regular and general alternation of day and night will come to an end with some abruptness; and the period of rotation be brought to coincide with the period of orbital revolution.

It is, of course, impossible in the existing state of knowledge in regard to planetary rotation to put forward the present argument as anything more than a vague conjecture. The observations of astronomers may in the future either strengthen the reasoning, or on the other hand, may completely upset it.   Our endeavour has been to show that the hypothesis that the rotation of planets has some direct relation to their individual condition as regards heat is consistent with the facts of their rotation in so far as these are now known.   What facts bearing on the subject the future may reveal, time alone can show.

THE DAILY BAROMETRIC TIDE.

THE DAILY BAROMETRIC TIDE.

ABOUT the year 1641, when the controversy, which culminated in the Civil War and the execution of King Charles the First, was raging throughout Britain, and its results still hanging in the balance, an insignificant circumstance occurred in Florence which was destined to prove of stupendous importance to the science of meteorology. This was merely the sinking of a pump to raise water for domestic purposes from a deep well. In Britain the throne was beginning to totter, families were divided, the nation was arming for the struggle; in Florence domestic concerns occupied the attention of the people. Yet in this case the peaceful sinking of the pump was more directly eventful to science than the conflict between King Charles and his Parliament.

The depth from which the Florentines desired, on this occasion, to raise the water was unusually great, and they were baffled in their efforts. Notwithstanding that every care was taken in fitting the piston and valves in the pump, the water would not rise to the surface. It was found that by no means could the water be made to rise higher in the pump than about thirty-two feet. Galileo, the illustrious astronomer, then in extreme old age, resided near Florence, and he was consulted. He was, however, unable to explain the phenomenon satisfactorily. Fortunately the subject came under the notice of his pupil, Torricelli, who recognized that a fundamental fact in natural science had been accidentally discovered. Torricelli meditated a long time, and at last, in 1643, came to the conclusion that the phenomenon could be explained only by attributing weight to the atmosphere. If then, water, through the weight of the atmosphere, was forced to rise in a pump to the height of about 32 feet by the withdrawal of the air in the pump, liquids heavier than water would be forced to rise a less distance. Torricelli took a glass tube about 4 feet in length, closed at one end and open at the other. He filled it with mercury, covered the open end with his finger, and turned the tube upside down. He then inserted the lower end in a basin of mercury, and withdrew his finger under the surface of the mercury in the basin. Holding the tube in a vertical position, he observed that the mercury in it sank a certain distance, and, after a few oscillations, settled at the height of about 30 inches above the surface of the mercury in the basin. On comparing the height of the column of mercury with the height of the column of water raised by the pump, he found these heights to be in an inverse ratio of the specific gravities of water and mercury. Observing also that the columns of water and mercury had no communication with the atmosphere at their upper extremities, but that they did indirectly communicate therewith at their lower extremities, he concluded that the two columns were suspended by the same cause, namely the weight of the atmosphere. For the first time in the history of the world, a column of the atmosphere had been consciously weighed in the

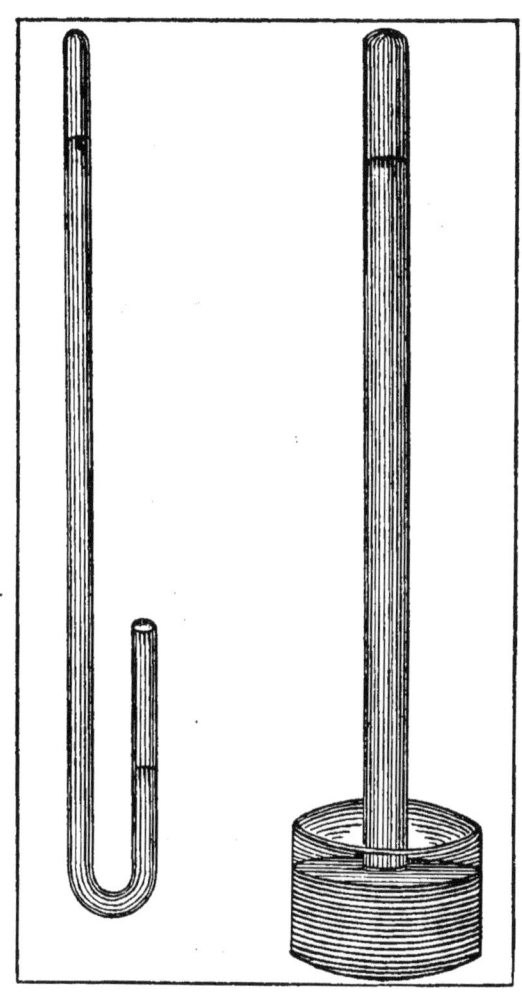

Siphon barometer and cistern barometer (in their essential and simplest forms).—p. 302,

balance. The discovery of the barometer had been made.

Pascal, the distinguished French philosopher, although only about 20 years of age when Torricelli, in 1643, made his brilliant discovery, was deeply interested; and, on his suggestion, a crucial test was carried out about the year 1648. If the liquid in the tube was really sustained by the weight of the atmosphere, it should follow that as the pressure of the atmosphere lessened, the height of the column would decrease, as the pressure of the atmosphere increased the height of the column would become greater. It was evident, also, that at the top of a mountain there could not be so great a thickness of atmosphere as at the base. Therefore the column of liquid in the tube would, if sustained by the pressure of the atmosphere be of greater height at the base of a mountain than at the summit. Pascal proposed that the experiment should be made, and it was arranged accordingly. Two tubes containing columns of mercury were conveyed to the mountain Puy de Dome, near Clermont, in Auvergne. At the base they both stood at 28 inches. One was left there and the other taken to the summit. As the observers ascended, the mercury in the tube sank gradually to 24·7 inches, and as they descended, the mercury as gradually rose again to its former elevation. This experiment left no room for doubt that the liquid was supported by the weight of the atmosphere.

The simple apparatus devised by Torricelli, consisting of a tube of mercury dipping into a cistern of the same metal, is, in fact, the barometer. Notwithstanding the endless varieties of shape, and the complications which have since been devised, it is doubtful whether in any essential particular we have improved on the simple form of instrument by means of which the atmospheric pressure was first discovered—except, indeed, in more careful construction and graduation of scale. The great principle discovered by Torricelli, that the weight of the atmosphere supports the liquid column in the tube, must always be recognized as one of the greatest discoveries in natural science—an outstanding event in the world's history.

As the barometer is simply an instrument for determining the weight or pressure of the atmosphere, the height of the barometer, of course, varies with the fluctuations in the pressure of the atmosphere. The height of the barometer has no direct relation to the state of the weather, and the words "Change," "Fair," "Stormy," etc., which are so frequently shown on the face of the barometer are, to some extent, calculated to mislead. It is only in so far as changes in the weather are coincident with changes in the atmospheric pressure that the barometer can be accepted as a reliable guide. If the barometer rises, we know that the pressure of the atmosphere overhead has increased; if the barometer falls we know that the pressure of the atmosphere overhead has decreased. Such increase or decrease in the pressure of the atmosphere may or may not be accompanied by characteristic weather changes. In order to ascertain the bearing of these barometric changes on changes of weather, we have to go outside the barometer, and take other matters into account, such as the direction of the wind, the season of the year, and local peculiarities. It is said, indeed, that a rise of the barometer between 9 a.m. and 3 p.m. is a fair indication of fine weather; while a fall of the barometer between 3 p.m. and 9 p.m. is a fair indication of rain.

The fluctuations in the height of the barometer are of two distinct kinds, regular or periodic, and irregular or non-periodic. The irregular fluctuations are those which we are accustomed to associate with changes in the meteorological conditions, and to accept as weather guides. Although the connection with the weather is only indirect, it is to be supposed that changes in atmospheric pressure will be associated with changes in the weather, which is, of course, purely an atmospheric manifestation. Thus, when due regard is paid to the special purpose of the barometer as an atmospheric balance, and to the relative phenomena to which we have referred, the use of the barometer as a weather prognosticator is of the highest value.

The regular, or periodic fluctuations of the barometer are amongst the most mysterious of meteorological phenomena, and their causes cannot be considered as being even yet fully understood. In fact the explanations which have been put forward, although in some cases by the highest authorities, have failed to secure general acceptance. Let us now consider these regular fluctuations in some detail.

The most noticeable of the regular fluctuations is that usually known as the diurnal barometric oscillation. It varies from an amplitude of perhaps three-twentieths of an inch in the tropics, to barely one-thousandth of an inch in the polar regions, its amount being thus according to latitude. It is said that in Britain the best times of the day for observing the barometer are 9 a.m. and 9 p.m., when on the average it stands higher, and 3 a.m.

and 3 p.m., when on the average it stands lower, than at any other times during the twenty-four hours. These times are, however, not absolutely constant throughout the year, but are subject to a certain degree of seasonal change. In tropical and temperate latitudes, it appears to be invariably the case that the barometer has an easily discoverable double oscillation in the day showing two maxima and two minima.

Dr. Alexander Buchan, the well-known meteorologist, states that

"Of the periodical variations of atmospheric pressure, the most marked is the daily variation, which, in tropical and sub-tropical regions, is one of the most regular of recurring phenomena. In high latitudes the diurnal oscillation is masked by the frequent fluctuations to which the pressure is subjected. If, however, hourly observations be regularly made for some time, the hourly oscillation will become apparent. The results show two maxima occurring from 9 to 11 a.m. and 9 to 11 p.m., and two minima occurring from 3 to 6 a.m. and 3 to 6 p.m."

From observations made at Calcutta during the six years from the beginning of 1857 to the end of 1862, it was found that the two daily maxima in the height of the barometer occur at about 9 or 10 o'clock in the forenoon, according to the season, and at about 10 o'clock in the evening in each season; and that the two daily barometric minima occur at about 3 o'clock in the morning, and 4 o'clock in the afternoon all the year round. At Vienna, the two daily maxima were found, by corresponding observations, to occur at about either 9 or 10 o'clock in the forenoon, and 10 or 11 o'clock in the evening; and the two daily minima at from 3 to 6 o'clock in the morning, and from 3 to 5 o'clock in the afternoon —the time of occurrence of the four extremes varying according to the season.

The maxima and minima recorded in the daily barometric swing in these two stations differed materially from each other in extent. In Calcutta, the morning maximum varied from ·079 of an inch above the daily mean, in January, to ·040 of an inch above the daily mean, in July—the corresponding figures for Vienna being ·018 and ·022· In Calcutta, again, the evening maximum varied from ·029 of an inch above the daily mean in July, to ·010 of an inch above the daily mean in January, while the corresponding figures for Vienna are ·009 and ·012 Thus the largest forenoon maximum occurs at Calcutta in January, and the smallest forenoon maximum in July; while exactly the converse is the case in Vienna. Again, the largest afternoon maximum occurs at Calcutta in July, and the smallest afternoon maximum in January, the records in Vienna being in this case also almost the converse.

As regards the daily minima of the barometric fluctuations, it was found in Calcutta that the morning fall of the barometer varied from ·026 of an inch below the daily mean, in October, to ·019 of an inch in July, the corresponding figures for Vienna being ·010 below the mean, and ·003 above it. The afternoon fall in Calcutta was much more appreciable than that of the early morning. It varied from ·071 of an inch in April to ·047 in October, while in Vienna it amounted in the same months to ·027 and ·015 respectively. The daily barometric oscillations which occur with such regularity in Calcutta and Vienna are typical, to a large extent, of

x

the oscillations which occur daily in every place throughout the tropical and temperate regions of the Globe. We therefore give the following Table showing the results of the six years' observations at these two stations.

TABLE SHOWING THE DAILY BAROMETRIC MAXIMA AND MINIMA, ACCORDING TO THE SEASON, AT CALCUTTA AND VIENNA RESPECTIVELY, WITH THE HOUR OF OCCURRENCE (FROM OBSERVATIONS MADE DURING THE SIX YEARS 1857.1862).

Calcutta. (Lat. 22° 33′ 47″ N.) (Long. 88° 23′ 34″ E.)			Vienna. (Lat. 48° 13′ N.) (Long. 16° 23′ E.)	
	Hour.	Variation from Daily mean.	Hour.	Variation from Daily mean.
Forenoon Minima.				
January.	3 a.m.	− ·023 of an inch	6 a.m.	− ·008 of an inch
April.	3 ,,	·020 ,,	5 ,,	− ·003 ,,
July.	3 ,,	·019 ,,	3 ,,	+ ·003 ,,
October.	3 ,,	·026 ,,	6 ,,	− ·010 ,,
Forenoon Maxima.				
January.	10 a.m.	+ ·079 of an inch	10 a.m.	+ ·018 of an inch
April.	9 ,,	·070 ,,	10 ,,	·021 ,,
July.	10 ,,	·040 ,,	9 ,,	·022 ,,
October.	9 ,,	·064 ,,	10 ,,	·020 ,,
Afternoon Minima.				
January.	4 p.m.	− ·053 of an inch	3 p.m.	− ·020 of an inch
April.	4 ,,	·071 ,,	5 ,,	·027 ,,
July.	4 ,,	·051 ,,	5 ,,	·028 ,,
October.	4 ,,	·047 ,,	4 ,,	·015 ,,
Afternoon Maxima.				
January.	10 p.m.	+ ·010 of an inch	10 p.m.	+ ·012 of an inch
April.	10 ,,	·016 ,,	11 ,,	·014 ,,
July.	10 ,,	·029 ,,	11 ,,	·009 ,,
October.	10 ,,	·018 ,,	10 ,,	·008 ,,

It is at once evident that the diurnal oscillations in

Calcutta are of much greater range than they are in Vienna. In fact, in only one instance does the extent of the phase in Vienna exceed that in Calcutta, this being the afternoon maximum for January, which, in Vienna, was found to exceed the daily mean by ·012 of an inch, while, in Calcutta, it is only ·010 of an inch. The Table shows conclusively, as regards these two towns—what is indeed undoubted—that the diurnal swing of the barometer is of greatest extent in the tropical latitudes, and becomes less noticeable as the latitude increases.

Dr. Buchan collected observations from upwards of 250 places in different parts of the Globe in order to obtain information on the subject of the daily barometric oscillations. From these observations certain facts have been elucidated as to the variation in the time of occurrence, and in the range, of the oscillations.

It was found that the forenoon maximum occurs in January from 9 a.m. to 10 a.m. as far north as latitude 50° N.; while in July it occurs from 9 a.m. to 10 a.m. only to about latitude 40° N. In January, in latitudes farther north than 50° N., the time of occurrence of the forenoon maximum was found to vary from 8 a.m. to noon; while in July, in latitudes farther north than 40° N., it was found to vary from 7 a.m. to 11 a.m., the last mentioned hour being general in north-western Europe.

The afternoon minimum was found to occur from 3 p.m. to 4 p.m. in January nearly everywhere over the Globe, a few exceptions occurring in north-western Europe—the extremes being 2 p.m. at Utrecht (about lat. 52° 5′ N. and long. 5° 7′ E.) and 6 p.m. at St Petersburg (lat. 59° 41′ N. and long. 30° 29′ E.) In July the

...which occur daily in every place
...tropical and temperate regions of the
...before give the following Table showing
...six years' observations at these two

...THE DAILY BAROMETRIC MAXIMA AND
...ING ON THE SEASON, AT CALCUTTA AND
...WITH THE HOUR OF OCCURRENCE
...MADE DURING THE SIX YEARS

...	Variation from Daily mean	Hour	(Lat. 48° 13′ N.) (Long. 16° 22′ E.) Variation from Daily mean
...	...008 of an inch	8 a.m.	− ·008 of an inch
...	·008 ,,	9 ,,	− ·008 ,,
...	·009 ,,	2 ,,	+ ·008 ,,
...	·008 ,,	3 ,,	− ·010 ,,
...	+ ·008 of an inch	10 a.m.	+ ·018 of an inch
...	·009 ,,	10 ,,	·021 ,,
...	·010 ,,	9 ,,	·022 ,,
...	·014 ,,	10 ,,	·020 ,,
...	·010 of an inch	3 p.m.	− ·028 of an inch
...	·021 ,,	5 ,,	·027 ,,
...	·021 ,,	5 ,,	·025 ,,
...	·047 ,,	4 ,,	·015 ,,
...	+ ·016 of an inch		
...	·016 ,,		
...	·029 ,,		
...	·018 ,,		

...bout the

Calcutta are of much greater range han they are in Vienna. In fact, in only one instance oes the extent of the phase in Vienna exceed that in Cacutta, this being the afternoon maximum for January, hich, in Vienna, was found to exceed the daily mean b 012 of an inch, while, in Calcutta, it is only 010 of an ch. The Table shows conclusively, as regards these o towns—what is indeed undoubted—that the diurnal swing of the barometer is of greatest extent in the opical latitudes, and becomes less noticeable as the latitude increases.

Dr. Buchan collected observations om upwards of 250 places in different parts of the lobe in order to obtain information on the subject of the aily barometric oscillations. From these observations rtain facts have been elucidated as to the variation n the time of occurrence, and in the range, of the oscillations.

It was found that the forenoon maximum occurs in January from 9 a.m. to 10 a.m. as far orth as latitude 50° N.; while in July it occurs from 9 a.m. to 10 a.m. only to about latitude 40° N. In January, in latitudes farther north than 50° N., the time of occurrence of the forenoon maximum was found to vary from 8 a.m. to noon; while in July, in latitudes farher north than 40° N., it was found to vary from 7 a.m to 11 a.m., the last mentioned hour being general i north-western

The afternoon m n was found o occur from 3 p.m. to 4 where over the Globe, a rn afternoon minimum was found to occur at from 3 p.m. to 4 p.m. only as far north as latitude 40° N., while in higher latitudes the time varied from 4 p.m. to 6 p.m.

Thus as far north as latitude 40° N. the forenoon maximum occurs all the year round from 9 a.m. to 10 a.m., and the afternoon minimum from 3 p.m. to 4 p.m. In higher latitudes—where the oscillatory movement is much smaller and observation consequently more difficult—it appears that the time of the occurrence varies, but, with few exceptions, not to any excessive extent.

Dr. Buchan considers that the decrease between the morning maximum, and the afternoon minimum is the most important matter in connection with the diurnal movement. The extent of this decrease is, generally speaking, greatest in the tropics, and diminishes as the latitude increases. It is greater on land than at sea, and increases greatly on proceeding inland. It is nearly always greater with a dry than with a moist atmosphere, and generally, but by no means always, it is greatest in the month of highest temperature and greatest dryness combined. In Central America the amplitude of the movement between the morning maximum and the afternoon minimum is about one-tenth of an inch, and this is exceeded in other parts of the tropics. In the tropical parts of the ocean the amplitude of the oscillation is less than it is on land, the difference being from ·020 to ·030 of an inch. The influence of the Mediterranean Sea in lessening the amount of the oscillation over all regions bordering it is very strongly marked.

It has been found that for fully 30° north and south

of the equator the time-line of maxima and minima, in the diurnal barometric oscillations, runs north and south, but that in higher latitudes the direction is changed, the changes being chiefly conspicuous as regards the forenoon maximum and the afternoon minimum, that is to, say the extremes which occur in the day-time as distinguished from the night-time. Thus, in the higher latitudes, a line indicating the same time of occurrence of the forenoon maximum or the afternoon minimum, would not extend northward, and southward, but would deviate to the west or east.

Dr. Buchan mentions that at 6 p.m. *of Greenwich mean time*, the afternoon minimum in June exists in 35° W. longitude (in which longitude the local time would be 3.40 p.m.) on all parts of a line extending from 30° N. latitude to 30° S. latitude. In the latitude of London (51° 30′ 48″ N.), this extreme occurs at the hour mentioned, not on the meridian 35° W. longitude, but at about 16° W. longitude—at which longitude the local time would be 4·56 p.m. On the other side of the equator, in the Falkland Islands (which lie between latitudes 51° S. and 52° 45′ S.), the occurrence of the same extreme at 6 p.m. *Greenwich mean time*, is found at about the meridian 60° W. longitude, where the local time would be 2 p.m. It follows that a line drawn southwards, indicating the positions at which this extreme occurs at exactly the same hour—6 p.m. of Greenwich mean time—would slope westward from 16° west longitude in the latitude of London to 35° W. longitude in latitude 30° N., and would then extend due southward to 30° S. latitude, and then again deviate westward until, in the Falkland Islands, its position would be 60° W. longitude.

Thus, in June the afternoon minimum occurs in local time about three hours earlier in the Falkland Islands than it does in the south-west of Ireland, indicating in a striking manner, according to Dr. Buchan, the influence of season on the phenomenon.

From hourly observations made in the month of June at the base, the top, and two intermediate points on Mount Washington, New Hampshire (about 44° 17' N. lat., and 71° 15' E. long.), it was found that the time of occurrence of the forenoon maximum at the base of the mountain (the base being 2,898 feet above sea-level), was 8 a.m.; at 4,059 feet it was 10 a.m.; at 5,533 feet, 11 a.m., and at the top (6,285 feet) it was noon. At the top of Mount Washington, it was found that the early morning fall of the barometer was much greater than that which occurred in the afternoon, the former being ·020 of an inch below the daily mean, and the latter only ·004 of an inch. At the base of the mountain the converse was the case, the early morning minimum being only ·006 of an inch below the daily mean, while the afternoon minimum was ·020 of an inch below the same level. Dr. Buchan compares the action of the barometer at the top of the mountain, in this respect, to what has been found to occur at insular stations, and its action at the base to what has been found to occur in continents.

The diurnal oscillations of the barometer take place over the sea as well as over the land surface. In fact while the nature of the surface may affect the extent, or time of occurrence, of the oscillations, it has no relation to the actual existence of the oscillatory movement. Dr. Buchan gives the following table of the daily movement of the barometer in the Pacific, as found by the

Challenger, from the 1st to the 12th of September 1875, in mean latitude 1° 8′ S., and longitude 150° 40′ W. The mean height of the barometer was 29·928 inches, and the signs *plus* (+) and *minus* (−) indicate respectively the elevation above, and the depression below the daily mean, which occurred at the hours specified.

DAILY MOVEMENT OF THE BAROMETER IN THE PACIFIC.
MEAN LAT. 1° 8′ S., LONG. 150° 40′ W.
(As found by the "Challenger" from 1st to 12th September 1875).

Hour.	Inch.	Hour.	Inch.	Hour.	Inch.
2 a.m.	− ·012	10 a.m.	+ ·032	6 p.m.	− ·028
4 ,,	− ·022	Noon.	+ ·006	8 ,,	+ ·004
6 ,,	+ ·003	2 p.m.	− ·043	10 ,,	+ ·013
8 ,,	+ ·028	4 ,,	− ·055	Midnight	+ ·012

The early morning minimum is evident as occurring about 4 a.m.; the forenoon maximum about 10 a.m.; the afternoon minimum about 4 p.m., and the afternoon maximum about 10 p.m. The amplitude of the range from the forenoon maximum (+ ·032) to the afternoon minimum (− ·055) which amounts to ·087 of an inch, is one of the most striking features in these oscillations, as also the rapidity of the fall between 10 a.m. and 2 p.m. The latter is believed to be characteristic of all observations within at least 12 degrees of the equator.

It has been found that, even though the mean atmospheric pressure may for a period be greater in temperate latitudes than in the tropics, the extent of the oscillations of the barometer near the equator are, nevertheless, greater than in the higher latitudes.

Dr. Julius Hann, a distinguished German meteorologist and an honorary member of the Royal Meteorological

Society, has given a great deal of attention to the regular barometic oscillations, and a translation of a valuable article by him on the subject is given in the *Quarterly Journal of the Royal Meteorological Society* for January 1899. Dr. Hann has come to the conclusion that, at the Earth's surface, *three* regular daily variations of the barometer occur, the first having a period of about 24 hours, the second of about 12 hours, and the third of about 8 hours. These are described, respectively, as the diurnal oscillation, the semi-diurnal oscillation and the ter-diurnal oscillation. The semi-diurnal oscillation is that with which we have been dealing, and which is specially noted by Dr. Buchan.

Dr. Hann states that the diurnal oscillation comes out much more decidedly on clear than on cloudy days. It has a small amplitude, or range from maximum to minimum, over the ocean, but a very great one over heated land-surfaces, and in mountain valleys, and is undoubtedly related to the diurnal range of temperature. It is subject to very great disturbance as to time and locality, and this disturbance is supposed to be connected with the fact that all meteorological phenomena exhibit principally a diurnal period, and consequently exert an influence on this barometric oscillation. All local modifications of meteorological conditions, as well as all temporary changes of weather, affect this movement. Thus the diurnal movement, identified by Dr. Hann, bears clear traces of all irregularities of meteorological phenomena, and, on this account, neighbouring localities may exhibit great differences both in amplitude and time of occurrence of extremes. "These things," according to Dr. Hann, "make it extraordinarily difficult to discover the constant portion of the diurnal barometric wave, which varies with latitude and season, among the interferences of the local diurnal barometric waves therewith."

The eight-hour oscillation, discovered by Dr. Hann, is of an excessively minute character. He gives its amplitude as lying between ·02 and ·05 millimetres—say ·0008 of an inch to ·002 of an inch. Dr. Hann was able to distinguish this movement, owing to the fact, as he mentions, "that hardly any meteorological phenomenon exists with a decided eight-hour period, so that the eight-hour barometric oscillation is almost undisturbed despite its slight amplitude."

The semi-diurnal oscillation is much more important than either of the other daily movements of the barometer referred to. Dr. Hann states that—

"The semi-diurnal wave shows itself at all places with a regularity quite unknown among other meteorological phenomena; along each parallel of latitude it comes out with almost uniform amplitude and phase epoch. The latter is (with reference to local time), almost quite constant up into high latitudes whereas the amplitudes steadily decrease with approach to the Pole."

He continues—

"It is very remarkable that the double daily oscillation is the principal phenomenon. It exhibits the greatest amplitudes and over equatorial oceans is almost the only one represented."

Dr. Hann emphasizes the fact that the semi-diurnal oscillation "is quite independent of the Earth's seasons, for it is the same in both hemispheres." He ascertained that "the principal maxima occur at the Equinoxes, the principal minimum in June and July, and a second, much smaller, minimum in December and January." In both hemispheres the amplitude or range of movement in this oscillation is greater when the Earth is nearest to the Sun—or in perihelion—than when it is in aphelion. The latter, indeed, coincides with the absolute minimum of amplitude. This movement is clearly recognizable day by day with almost absolute symmetry in low latitudes, but, in the higher latitudes, it is rarely traceable by direct observation. Dr. Hann states that it is not affected in either its amplitude or phase-time by the weather, and expresses the opinion that by the constancy of its elements along each parallel of latitude, and by its regular variation according to season and latitude, the semi-diurnal oscillation reminds us of the behaviour of cosmical phenomena.

As to the causes of the daily movements of the barometer, there is much uncertainty. Dr. Buchan, indeed, remarks that—

"Though the diurnal barometric oscillations are among the best marked of meteorological phenomena, at least in tropical and sub-tropical regions, yet none of these phenomena, except perhaps the electrical, could be named, respecting whose geographical distribution so little is really known, whether as regards the amount of variation, the hour of occurrence of the critical phases, or, particularly, the physical causes on which the observed differences depend."

Professor Cleveland Abbe, the Meteorologist of the United States Weather Bureau, expresses the opinion that—

"The dynamics and physics of the atmosphere have not yet been explored so exhaustively as to explain fully these three systematic barometric variations, but neither have we as yet any necessity for appealing to some unknown cosmic action as a possible cause of their existence."

In the opinion of Dr. Buchan—

"Much which has been written regarding these fluctuations, and in explanation of them, does not rest on facts; and nearly everything yet requires to be done in the way of collecting data towards the representation and explanation of the daily oscillations of atmospheric pressure, which are, as regards two-thirds of the Globe, perhaps, the most regular of recurring phenomena, and an explanation of which cannot but throw much light on many of the more important and difficult problems of the atmosphere."

Each of these opinions was, however, written some years ago, and, even in the articles in which they are severally expressed, each of the high authorities whom we have quoted, goes some way in the direction of furnishing an explanation of the occurrence of the phenomena dealt with, and adds somewhat to the data necessary to allow of an explanatory theory being fairly weighed.

Dr. Buchan concludes, from the observations he has collected, that, whatever be the cause or causes to which the daily barometric oscillation is due, the absolute amount is largely dependent on comparatively local influences. Dr. Hann, however, in dealing with the same phenomenon—the double daily oscillation—states that local disturbances of the magnitude of the movement and of the time of its occurrence, are small in general, and have not been much investigated.

It is conjectured that the primary cause of the different regular barometrical oscillations is to be found in the action of solar heat upon the illuminated hemisphere, and the many consequences that result therefrom. According to Professor Abbe, "Among the many methods of action that have been studied or suggested in connection with the barometric variations, the most important of all is the so-called tidal wave of pressure, due to temperature."

Dr. Buchan, in dealing solely with the semi-diurnal oscillation, briefly indicates, but partially dissents from, an explanation which had been suggested of the phenomenon as follows—

"Since the two maxima of daily pressure occur when the temperature is about the mean of the day, and the two minima when it is at its highest and lowest respectively, there is thus suggested a connection between the daily barometric oscillations and the daily march of temperature; and similarly a connection with the daily march of the amount of vapour and humidity of the air. The view entertained by many of the causes of the daily oscillations may be thus stated:—The *forenoon maximum* is conceived to be due to the rapidly increasing temperature, and the rapid evaporation owing to the great dryness of the air at this time of the day, and to the increased elasticity of the lowermost stratum of air which results therefrom, until a steady ascending current has set in. As the day advances the vapour becomes more equally diffused upwards through the air, an ascending current, more or less strong and steady, is set in motion, a diminution of elasticity follows, and the pressure falls to the *afternoon minimum*. From this point the temperature declines, a system of descending currents sets in, and the air of the lowermost stratum approaches more nearly the point of saturation, and from the increased elasticity, the pressure rises to the *evening maximum*. As the deposition of dew proceeds, and the fall of temperature and consequent downward movement of the air are arrested, the elasticity is again diminished, and pressure falls to the *morning minimum."*

Dr. Buchan adds with reference to this explanation—

"Since the view propounded some years ago, that if the elastic force of vapour be subtracted from the whole pressure, what remains will show only one daily maximum and minimum, has not been confirmed by observation, it follows that the above explanation is quite insufficient to account for the phenomena; indeed, the view can be regarded in no other light than simply as a tentative hypothesis."

Dr. Buchan thinks that there can be no doubt from the facts of the case, so far as they have been disclosed, that the daily barometric oscillations originate with the Sun, but that more than the Sun's influence, as exerted on the diurnal march of the temperature and humidity of the atmosphere, is concerned in their production. He is satisfied that, whatever the originating cause or causes, the effects are enormously modified by the distribution of land and water, by the wind, and by the humidity of the atmosphere.

In a later article—on "Meteorology"—which appears in the *Encyclopædia Britannica,* Dr. Buchan indicates that, in his view, the cause of the semi-diurnal oscillation lies in the solar heating of the Earth's atmosphere, section by section, which results from the Earth's rotation—and the consequent aerial currents.

"Owing to the rapidity of the diurnal heating of the atmosphere by the Sun through its whole height, some time elapses before the higher expansive force called into play by the increase of temperature can counteract the vertical and lateral resistance it meets from the inertia and viscosity of the air.. Till this resistance is overcome, the barometer continues to rise, not because the mass of atmosphere overhead is increased, but because a higher temperature has increased the tension or pressure. When the resistance has been overcome, an ascending current of the warm air sets in, the tension begins to be reduced, and the barometer falls, and continues to fall till the afternoon minimum is reached. Thus the forenoon maximum and afternoon minimum are simply a temperature effect, the amplitude of the oscillation being determined by latitude, the quantity of aqueous vapour overhead, and the Sun's place in the sky."

The evening maximum and early morning minimum are accounted for as follows—

" When the daily maximum temperature is past . . . the air becomes more condensed in the lower strata, and pressure consequently at great heights is lowered. Owing to this lower pressure in the upper regions of the air, the ascending current which rises from the longitudes where at the time the afternoon pressure is low, flows back to eastward, thus increasing the pressure over those longitudes where the temperature is now falling. This atmospheric quasi-tidal movement occasions the p.m. increase of pressure, which reaches the maximum from 9 p.m. to midnight, according to latitude and geographical position. This maximum is therefore caused by accessions to the mass of the atmosphere overhead, contributed by the ascending currents from the longitudes of the afternoon low pressure immediately to westward. As midnight and the early hours of the morning advance, these contributions . . . cease . . . and pressure continues steadily to fall. . . . Radiation towards the cold regions of space takes place, not only from the surface of the Globe, but also directly from the molecules of the air, and its aqueous vapour. The effect of this . . . cooling of the atmosphere . . . is necessarily a diminution of its tension. Since this takes place at a more rapid rate than can be compensated for by any . . . tidal movement of the atmosphere from the regions adjoining, owing to the inertia and viscosity of the air, pressure continues to fall to the morning minimum. This minimum is thus due not to the removal of any of the mass of air overhead, as happens in the case of the afternoon minimum, but to a reduction of the tension or pressure of the air, consequent upon a reduction in the temperature through radiation from the aerial molecules towards the cold regions of space."

Lord Kelvin, in 1882, in a contribution to the *Proceedings of the Royal Society of Edinburgh*, suggested that "the free oscillation produced by a relatively small amount of tide-producing force will have an amplitude that is larger for the half-day term than for the whole-day term"—that is, the free oscillation will be larger for a twelve-hour period than for a twenty-four-hour period. The tide-producing force in the atmosphere is believed to be the increase of temperature resulting from solar heat. This has, of course, a twenty-four-hour period, so that a diurnal wave in the atmosphere with that period is thereby occasioned. This gives rise to, and reinforces, natural oscillations of the atmosphere, whereby what are called "free waves" are set up, as contrasted with the "forced waves" directly produced by the diurnal march of temperature. Lord Kelvin suggested that the period of these "free waves," or the free oscillation of the terrestrial atmosphere, was really about 12 hours. Dr. Hann goes minutely into this matter, and supports the proposition of Lord Kelvin. Professor Abbe writes regarding the subject in the following terms—

"The process by which a diurnal temperature wave produces a semi-diurnal pressure oscillation . . . may be stated as follows :—The diurnal temperature wave, having a twenty-four hours' period, is the generating force of a diurnal pressure tide, which is essentially a forced and small oscillation. The natural period of the free waves in the atmosphere agrees much more nearly with twelve than with twenty-four hours. In so far as the forced and the free waves reinforce each other, the semi-diurnal waves are reinforced far more than the other, so that a very small semi-diurnal term in the temperature oscillations will produce a pressure oscillation two or three times as large as the same term would in the diurnal period. These reinforcements, however, depend upon the elastic pressure within the atmosphere, just as does the velocity of sound. If the prevailing barometric pressures were slightly increased, the adjustment of the twelve-hours' free wave of pressure to the forced wave of temperature could be so perfect that the barometric wave would increase to an indefinite extent. For the actual temperatures, the periodicity of the free wave is about thirteen hours, or somewhat longer than the forced wave of temperature, so that the barometric oscillation does not become excessive."

Dr. Hann, while strongly supporting the suggested explanation, states that he has received, from some of his colleagues, certain objections to Lord Kelvin's hypothesis of the origin of the double daily barometric wave. He considers these of such importance that he summarizes them—

"First, it is objected that it is questionable if the march of temperature in the upper strata is precisely of the same character as in the lower, on which latter the hypothesis is based ; and secondly, that the double daily temperature wave is only a result of calculation, a mathematical fiction, which cannot serve in the explanation of a real phenomenon. There is, in reality, no daily variation of temperature with two maxima and minima; and if we, in spite of this, obtain a double daily temperature wave, because we insist on representing the daily march of temperature by a series of sines, this forced mathematical form can never serve to explain an observed phenomenon, whereas for this some real natural process must be sought for as a cause."

Dr. Hann, in supporting the hypothesis, combats these objections.

The present position of the scientific world, in regard to the question of how the semi-daily barometric oscillations can be explained, is fairly indicated by the hypotheses to which we have referred; and by the objections noted by Dr. Hann. Many are, doubtless, satisfied by one or other of the explanations suggested, while some are not satisfied. On the whole, such a state of matters is to be welcomed, as conducing to further investigation and more convincing proof.

As regards the twenty-four-hours' oscillation recognized by Dr. Hann, it appears to be generally agreed that it can be explained by the diurnal temperature wave. It is a movement of great irregularity, being, as we have seen, much more decided on clear than on cloudy days, and having a small amplitude over the ocean, but a very great amplitude over heated land-surfaces and in mountain valleys. Dr. Hann's statement that "it undoubtedly proves itself by its striking local peculiarities to be related to the diurnal range of temperature" is beyond question.

The eight-hours' movement discovered by Dr. Hann is almost infinitesimal in amplitude, and it requires further

TIDE

investigation. It is not specially referred to in the various explanatory hypotheses which have been put forward.

The semi-diurnal oscillation is certainly the movement of greatest importance and interest, and the movement which has received the most attention. Nearly every writer on the subject has made reference to this oscillation as being of a very remarkable character. Dr. Hann himself expresses the opinion that any relation between this pressure oscillation and the diurnal range of temperature is very obscure. As we have noticed, even the high authority of the physicists, who have come to the conclusion that the phenomenon can be explained by the free oscillations of the atmosphere as influenced by the diurnal variations in temperature, has not satisfied all enquirers. There are still some who think that in order to satisfactorily account for a real phenomenon, like the daily recurring double maxima and minima of the barometer, "some real natural process" must exist other than the daily temperature wave.

It seems possible that an explanation of the nature required by these objectors—that is to say, a real natural process—may actually exist. The actual movements of the Earth itself may possibly so effect the surrounding atmosphere, as to produce a semi-diurnal increase and decrease in the mass of the atmosphere overhead, and a consequent rise and fall of the barometer. Let us see whether there is any reasonable ground for such a statement.

In the first place we should recall that the barometer is simply an instrument for measuring the weight or pressure of the atmosphere. So far as we are able to judge the barometer rises and falls in conformity with an actual variation in the mass of the atmosphere overhead. When the atmospheric pressure is becoming greater, and the mercury in the cistern is being pressed down and forced into the tube, and a rise in the barometer consequently occurs, there is reason to believe that there is actually a greater mass of atmosphere over the barometer than there previously was. When, on the other hand, the atmospheric pressure is becoming less, and the mercury in the cistern rises through the pressure not being sufficient to support so much mercury in the tube, and the barometer therefore falls, there is reason to believe that there is actually a less mass of atmosphere over the barometer than previously. As the atmosphere is not really confined there appears to be reasonable ground for concluding that the changes in the barometer coincide with actual physical increases or decreases in the mass of the atmosphere overhead, and not merely with changes in the tension of the atmosphere.

Now, we know that if a current of the atmosphere flows from a part of the Earth where the axial velocity is excessive towards a part of the Earth where it is not so great, as, for instance, from the equator northwards or southwards, it retains for a time a greater axial velocity than pertains to the part of the Earth towards which it has flowed. As the Earth moves eastward, an atmospheric current flowing from the equator northward is felt as a south-west wind, its southerly character arising from the fact that it comes from a part of the Earth farther south than the region at which it is felt, and its westerly character from the fact that it comes from a part of the Earth which is rotating more rapidly

from west to east than is the region at which it is felt.
Similarly, an atmospheric current advancing southward
towards a lower latitude, is felt as a north-east wind, the
northerly character arising from the fact that it proceeds
from a higher latitude, and its easterly character from
the fact that it proceeds from a region where the
rotatory velocity of the Earth from west to east is not so
rapid.   The region which it has reached is turning
eastwards more rapidly than the atmosphere, and thus is
dashing through the atmosphere, which consequently
appears to blow from the east.   This clearly shows that
when any portion of the atmosphere changes its position
in relation to the surface of the Earth, *time* is necessary
to alter the velocity which it had acquired in its former
situation, to the actual velocity of the surface of the
Earth in the new situation. The attachment of the
atmosphere to the Earth being comparatively loose, any
change in the velocity of the surface with which, for the
time, it is in contact, or to which it is nearest, as compared with the velocity of the surface with which it was
previously in contact, or to which it was nearest, is only
by degrees taken up by the atmosphere, and until it is
taken up, the atmospheric movement is, to a gradually
lessening extent, characteristic of the surface movement
of the region from which it has proceeded. This may,
indeed, be accepted as almost an axiomatic statement in
relation to the atmosphere.

If, then, any portion of the atmosphere retains for a
time in a new situation, and only gradually loses, the
velocity communicated to it by the region of the Earth
from which it has passed, rather than the velocity of
the region which it now occupies, it cannot be doubted

that *if the situation of the atmosphere is unchanged but the velocity of the region which it occupies changes,* the atmosphere will for a time retain the previous velocity, and will only gradually accommodate itself to the altered conditions. So far as the atmosphere is concerned, such a change is exactly analogous to a change of situation. It has been, we may suppose, accompanying the region which it occupies, in its movements through space, and accommodating itself to the velocity of the region. If the velocity alters, the atmosphere, although unchanged, will clearly require time to adjust itself to the new conditions.

Is there any reason to suppose that the velocity of the same locality passes through certain changes in the course of the Earth's daily rotation? Let us see.

The daily movement of the Earth has a double character. The Earth each day rotates on its axis, and moves through an arc of its orbit. The axial velocity does not, of course, vary throughout the day. Once every 23 hours, 56 minutes, 4 seconds—the sidereal day—the Earth performs an axial rotation, and the portion of the rotation accomplished during each minute of that time is the same. The axial velocity may possibly change slowly, in the course of ages, but it is certainly as nearly as possible constant on each succeeding day. Therefore, in relation to the axial velocity, there is no possibility of a variation in movement being imparted to any portion of the atmosphere which remains in contact with the same locality.

The orbital velocity of the Earth undoubtedly changes with the increase or diminution of the Earth's distance from the Sun. When the Earth is in perihelion, or at its nearest distance to the Sun, the orbital velocity is much greater than when it is in aphelion, or farthest away from the Sun. As the distance from the Sun is every moment varying to a slight extent, there must every moment be a slight variation in the Earth's orbital velocity. Still, it is evident that the variation in the velocity thus produced is not of a *daily recurring* character. It has no likeness to a daily, or semi-daily, oscillation. The velocity slowly increases day by day and hour by hour, as the separation from the Sun lessens, and slowly decreases, in a similar manner, as the separation from the Sun becomes greater. Thus, the oscillation of the orbital velocity has a yearly period —a period coincident with that of the orbital revolution —and has no movement which can be considered as an oscillation in relation to the daily progress of the Earth.

It is a remarkable fact, however, that these two movements—the axial rotation and the daily orbital progress—result unitedly in every part of the Earth's surface (except, indeed, the Poles, where there is no rotational velocity) performing daily a journey of varying velocity, a journey in which the velocity rises to a maximum and falls to a minimum in the course of every twenty-four hours, the chief effect of the variation being at, or in the neighbourhood of, the equator, and the least effect in the neighbourhood of the Poles.

It is evident that, once in each rotation, every geographical position—at any rate in the tropical latitudes —faces more or less directly the direction of orbital progress—the direction in which the Earth is moving in its orbital revolution. About six hours afterwards, the same position faces the Sun, that is to say, the Sun is on the meridian. The position is then turned towards the central portion of the orbital circuit. Six hours later the same part is turned towards the portion of the orbit which the Earth has just passed along. We are, as it were, looking backward. After a further interval of about six hours, the same meridian is turned directly away from the Sun. The part of the heavens directly overhead lies altogether outside our orbit, our view is exactly in the direction opposite to the central portion

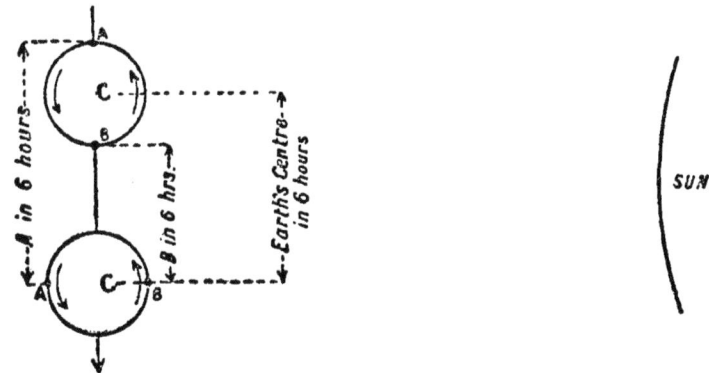

Diagram showing the varying surface velocity of the Earth in the course of the diurnal rotation,

of the orbital circuit. It is clear that when our position in the axial rotation lies between the part facing the portion of the orbit which the Earth has just passed and the part facing the direction of orbital progression—when we are rotating, as we might say, from the rear part of the Earth towards the foremost part—we are then moving axially in the same direction as the Earth is advancing in its orbital course. When, on the other hand, we are rotating, as we might say, from the foremost part of the Earth towards the rear part we are moving axially in the opposite direction from that in which the Earth is advancing in its orbital course.

In the former case we are, locally, moving in space at a greater rate than the Earth as a whole is moving. In the latter case, we are, locally, moving in space at a slower rate than the Earth as a whole is moving.

The orbital velocity of the Earth, as a whole, is about $18\frac{1}{2}$ miles a second, or 1,110 miles a minute. This is about the mean orbital velocity of the centre of the Earth. The axial velocity at the equator is about $17\frac{1}{3}$ miles a minute. Thus, when a position on the equator faces the direction of orbital progress, it is moving through space at the same velocity as the centre of the Earth—about 1,110 miles a minute. This also occurs when the position faces the part of the orbit over which the Earth has just passed.

A very different state of matters, however, arises in the intermediate positions. When a point on the equator is turned directly away from the centre of the orbit, as is the case about midnight, the full amount of its axial velocity is in the direction of orbital progress. The axial velocity of $17\frac{1}{3}$ miles a minute falls to be added to the orbital velocity, of, say, 1,110 miles a minute, in order to arrive at the local velocity. The point is moving through space at the rate of about $1,127\frac{1}{3}$ miles a minute. Again, when the same point has arrived at the opposite part of the axial movement, when it is turned towards the centre of the orbit—as occurs about noon daily—the full extent of the axial velocity is in the opposite direction from that of the orbital progress, so that the axial velocity of $17\frac{1}{3}$ miles a minute then falls to be deducted from the orbital velocity of about

1,110 miles a minute, in order to give us the local velocity, which therefore is about 1,092⅔ miles a minute.

Therefore, in one part of the axial rotation the local velocity at the equator is about 34⅔ miles a minute greater than it is at another part. The period of least velocity occurs about noon, when the position is facing the centre of the orbit. The velocity then gradually rises for the next twelve hours. About 6 p.m.—when the position is facing the part of the orbit over which the Earth has just passed—the velocity is at the mean, and at midnight the maximum is attained. The velocity then begins to lessen, and for the ensuing twelve hours it gradually falls. The mean velocity again occurs about 6 a.m., and at noon, the velocity is once again at the minimum.

Let us now confine our attention to a position at the equator, and see how this gradual variation in the velocity would be likely to affect the atmosphere resting on the Earth's surface. We may suppose our observation to commence at midnight when the velocity is at its maximum. In six hours the rate of motion will have fallen by about 17⅓ miles a minute. To begin with, the rate of motion is about 1,127⅓ miles a minute. At the end of six hours it is about 1,110 miles a minute, which is the mean rate of the Earth's orbital progress. The Earth, as a whole, will in the course of these six hours, have moved forward in its orbit by 399,600 miles, while the part of the Earth which we are observing will have moved forward one-half of the Earth's diameter more than that distance. As the diameter at the equator is about 7,926 miles in length, the point of observation will, in the time mentioned, have moved forward 403,563 miles. This gives an average velocity of 1,121 miles a minute, or 11 miles a minute in excess of the mean velocity of the Earth as a whole.

It may readily be supposed that the neighbouring atmosphere having—more particularly in its higher levels—received the impetus of the highest velocity, which prevailed at the time our observation is supposed to commence, and only gradually becoming subject to the decreasing velocity, will to a certain extent dash along ahead of the position with which it was in contact. The extent of the atmosphere resting on the place of observation will be thereby reduced, and the barometer will consequently fall. Thus, in this first quadrant of the equatorial circumference, lying between midnight and 6 a.m., a barometric minimum might reasonably occur.

From 6 a.m. till noon the local velocity still falls. We start with a velocity of about 1,110 miles a minute at about 6 a.m., and at noon our velocity is about $1,092\frac{2}{3}$ miles a minute. During that time the Earth, as a whole, moves forward in its orbit 399,600 miles, but our position moves forward one-half of the Earth's equatorial diameter less than that distance. Our change of position is thus 395,637 miles in the six hours, being somewhat less on the average than 1,099 miles a minute—or about 11 miles a minute less than the mean velocity of the Earth as a whole. It seems not at all improbable that in this quadrant the atmosphere propelled forward from the preceding quadrant by the excessive velocity will again be overtaken. As the velocity is now approaching the minimum, it is evident that no forward propulsion of the atmosphere can occur. Its natural inertia will have due effect, and thus our point of observation may come up with the portion of the atmosphere propelled forward from its own region, and at the same time be the recipient of the atmospheric accretions which will flow forward from the succeeding quadrant. Thus, in this quadrant, a superfluity of the atmosphere may occur, with consequent increase in atmospheric pressure and a barometric maximum.

The third quadrant extends from about noon to about 6 p.m. It begins with the velocity at the minimum, and ends with it at the mean, the average velocity in its course being about 11 miles a minute less than the average orbital velocity of the Earth as a whole. In this quadrant, our point of observation has a steadily increasing velocity. It would therefore seem likely that the surface, as it gains speed, will leave the atmosphere to some extent behind it—especially in the higher levels. This would induce a lessening height of atmosphere, and, consequently, a lower pressure with a barometric minimum. This minimum would arise in a converse manner to that in the first quadrant. Between midnight and 6 a.m., the atmosphere will have a higher velocity than the Earth, and an atmospheric depression will arise through the atmosphere rushing on ahead. Between noon and 6 p.m., the Earth will have a higher velocity than the atmosphere, and an atmospheric depression will arise through the earth's surface rushing on ahead.

In the fourth quadrant, our position of observation continues steadily to gain speed. The velocity commences at the mean, and ends at the maximum. This is the quadrant situated between 6 p.m. and midnight.

Through this increasing velocity, it would seem that portions of the atmosphere will constantly, as in the preceding quadrant, be left behind. This being so, a position must exist where these detached portions of the atmosphere will be, as it were, heaped up, so that the surface as it passes onward will, by degrees, come under a great mass of it. Each region as it advances will be the recipient of portions of the atmosphere left behind in the onrush of the preceding regions with their increasing speed. The atmospheric pressure will thus be increased, and a barometric maximum induced.

On this reasoning, we find that there are two positions of mean velocity which occur about 6 a.m. and 6 p.m. At each of these positions there is on the one side an elevation, and on the other side a depression of the atmosphere. These positions of elevation and depression might be expected generally to exist about midway between (1) the position of maximum velocity and the two positions of mean velocity, and (2) the position of minimum velocity and the two positions of mean velocity. As the position of maximum velocity is midnight, we have as the intermediate positions, having midnight as the centre, 3 a.m. and 9 p.m. As the position of minimum velocity is noon, we have similarly 9 a.m. and 3 p.m. as the intermediate positions.

The minimum before, and the maximum after, 6 a.m. would both be occasioned by the higher velocity of the atmosphere in the constantly decreasing velocity of the surface. The minimum before, and the maximum after, 6 p.m. would both be occasioned by the slower velocity of the atmosphere in the constantly increasing velocity of the surface.

With separation, and increase of distance, from the equator, the variation in velocity will gradually become less marked, and consequently the effect of the variation on the atmosphere will gradually lessen. The circumference of the Earth at the equator, or, as we may put it, the *length* of the equator, is about 25,000 miles, so that every position on the equator is caused, merely by the Earth's rotation, to pass through about 25,000 miles in about twenty-four hours, being, as we have seen, a velocity of about $17\frac{1}{3}$ miles every minute. The length of the circle forming, for instance, the 50th parallel of latitude is, however, not 25,000 miles, but only about 16,000 miles, so that every part of that parallel passes, by the Earth's axial movement, through only about 16,000 miles in each diurnal rotation, being a velocity of about $11\frac{1}{9}$ miles a minute. Thus, as the mean velocity of the Earth's orbital movement is about 1,110 miles a minute, each position in latitude 50° will have a maximum velocity, at midnight, of about $1,121\frac{1}{9}$ miles, and a minimum velocity, at noon, of about $1098\frac{8}{9}$ miles; the difference between the maximum and the minimum being thus about $22\frac{2}{9}$ miles, as compared with $34\frac{2}{3}$ miles at the equator. Similarly, between the equator and latitude 50°, the difference between the maximum and the minimum velocities will be greater than it is at latitude 50°, and less than it is at the equator; and between latitude 50° and the Poles, the difference will be less than it is at the 50th parallel. At the equator the difference is greatest, and, as the latitude increases, the difference proportionally lessens. Consequently, the quantity of the atmospheric ocean, which is caused to rush onward when the local surface velocity is falling, will neither be so great, nor the force of the propulsion so strong, at a distance from the equator as at the equator. In like manner, when the surface velocity is increasing, the amount of the atmosphere left behind in the onward rush of the Earth will be greater at the equator than elsewhere. There will, in fact, be a steady decline in the effects produced according to the distance of separation from the equator.

It will be seen that such an explanation of the semi-diurnal barometric oscillation would suggest why a difference occurs in the time of maxima and minima at different surface levels on a mountain, the atmospheric currents being influenced to a varying extent by the interposition of the mountain.

This explanation of the semi-diurnal swing would, if correct, seem to indicate that, between the times of the early morning minimum and the forenoon maximum, the wind should have generally a westerly character, while, between the times of the afternoon minimum and the evening maximum, its prevailing character should be somewhat easterly. In the early morning, owing to the falling surface velocity, the atmosphere rushes ahead, and, as the rotational movement of the Earth is from west to east, the air, as it rushes forward, will seem to move from the west. In the evening again, owing to the increasing surface velocity, the atmosphere falls behind, so that it will then appear to move from the east. It is, probably, only in the higher levels that such atmospheric currents would be noticeable. In fact, at the surface level their directions (according to the usual movements of the atmosphere in its efforts to regain equilibrium), might be absolutely different; just as an inflowing current at the surface is accompanied by an outflow in the higher levels, and *vice versa*.

We have noticed that, according to Dr. Buchan, the forenoon maximum occurs between 9 a.m. and 10 a.m. in the tropical and lower temperate regions, while, in the higher latitudes, the time of occurrence varies from 7 a.m. till noon. It would thus appear that, in every case, the forenoon maximum occurs in what we have dealt with as the "second quadrant," that is the region lying between 6 a.m. and noon. The afternoon minimum occurs between 3 p.m. and 4 p.m. in the tropical and lower temperate latitudes, while, in the higher latitudes, the time of occurrence varies from 3 p.m. to 6 p.m. Thus this phase appears invariably to occur in the "third quadrant," which lies between noon and 6 p.m.

As we have seen, Dr. Hann draws special attention to the fact that this oscillation is, in its amplitude, independent of the Earth's seasons, being the same in both hemispheres. He found that the principal maxima occur at the equinoxes, and the principal minima at the solstices. In both hemispheres, the amplitude of the oscillation is greater at perihelion than at aphelion, being, as we have mentioned, actually smallest at the latter epoch. Can these circumstances be accounted for consistently with the supposition that the daily changes in local velocity are the primary cause of this phenomenon?

This raises the question whether (1) the changing relation between the inclination of the Earth's axis and the direction of orbital progress, and (2) the varying orbital velocity, may exercise an influence on the magnitude of the extremes, and the amplitude of the oscillations.

At the equinoxes, one or other extremity of the Earth's
axis is inclined towards the forward part of the Earth in
the orbit; while, at the solstices, it is inclined towards
the Sun. At the equinoxes, the part of the Earth which
projects farthest forward, as the Earth speeds onward in
its orbital course, is a point about $23\frac{1}{2}$ degrees distant
from the equator (being about that distance north of the
equator at the autumnal equinox, and south of it at the
vernal equinox); while, at the solstices, the corresponding
part of the Earth is situated on the equator itself. At
the equinoxes, the Earth is practically at its mean
distance from the Sun; while, at the solstices, the Earth
is practically at one or other of the extremes of its
distance from the Sun—at about the greatest extreme
at the June solstice, and at about the smallest extreme,
or least distance, at the December solstice. At the
equinoxes, the Sun is on the equator; while at the
solstices it is most distant from the equator.

Setting aside for the moment the variation in the
distance of the Earth from the Sun, we find that the
local geographical relation to the orbital path is absolutely
different at the equinoxes, when the principal maxima,
or greatest atmospheric accumulations, occur, from what
it is at the solstices, when the principal minima, or
greatest atmospheric depressions, occur. We also find
that a very similar relation to the orbital path exists at
both solstices, and that this can also be said of both
equinoxes. There are, no doubt, geographical differences,
but, taking the Earth as a whole, the similarities at both
solstices, as compared with each other, and the similarities
at both equinoxes, as compared with each other, are no
less striking than the diversities at the solstices as

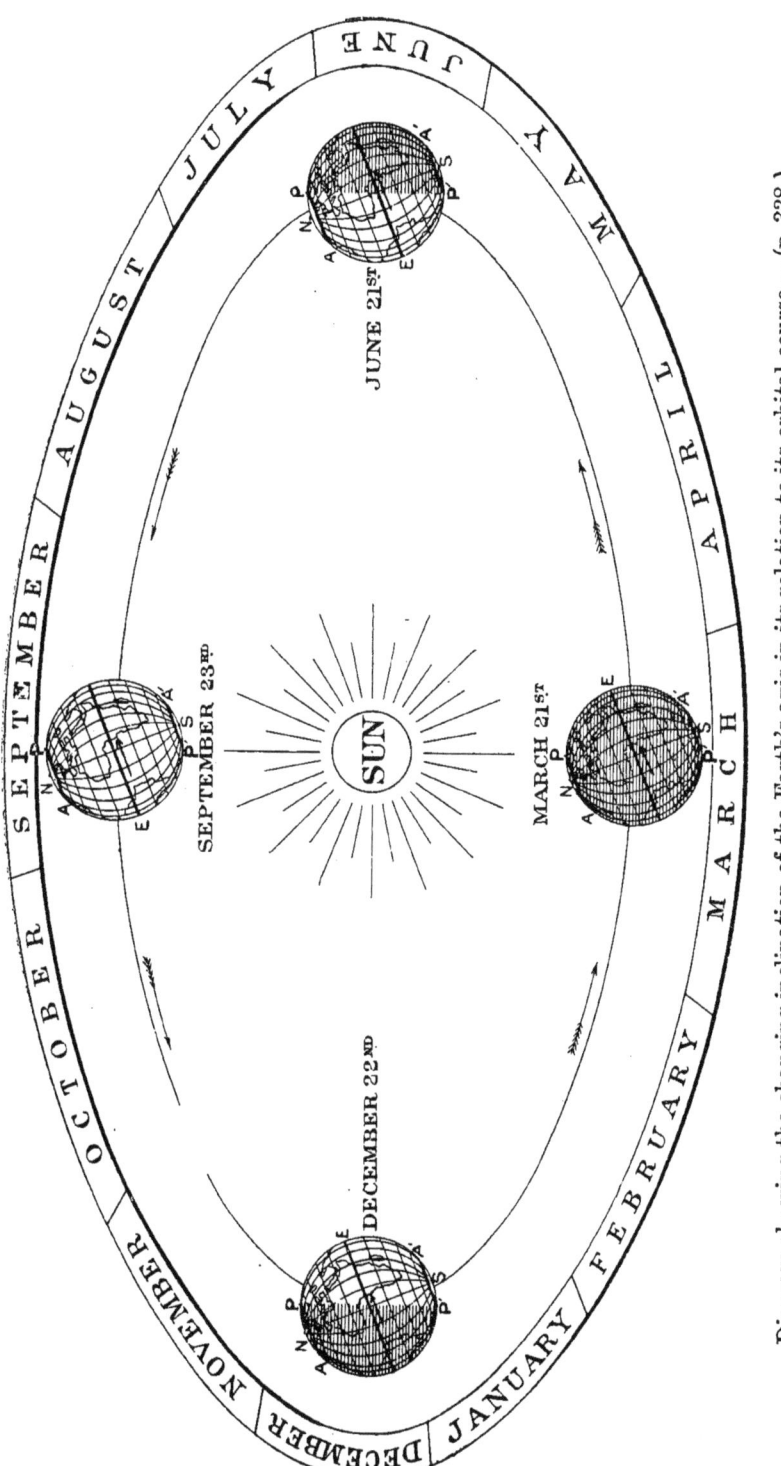

Diagram showing the changing inclination of the Earth's axis in its relation to its orbital course.—(p. 338.)

AP, A'P', Arctic and Antarctic Circles; P,P', Orbital Poles; N,S, Geographical Poles; E, Geographical Equator.

compared with the equinoxes, and at the equinoxes as compared with the solstices.

The points in which the positions at the solstices so strongly resemble each other, and differ from the positions at the equinoxes, and the points in which the positions at the equinoxes so strongly resemble each other, and differ from the positions at the solstices, are, clearly, the deciding points as regards variation in the character of the extreme, as distinguished from variation in amplitude, or in the magnitude of the extreme. They secure, when in a certain relation—as at the solstices—an increase of the atmospheric depressions; and when in another relation—as at the equinoxes—an increase of the atmospheric accumulations. In the former case, a greater amount of atmosphere would seem to be displaced to produce the greater depression, but it is, apparently, more dispersed, and therefore does not give rise to an excessive elevation. In the latter case, a smaller quantity of the atmosphere would seem to be displaced, but, apparently, in a more equable manner, and, consequently, its displacement gives rise to a more evident accumulation.

The various points of difference which occur at one solstice as compared with the other, result in differences in the magnitude of the extreme, as distinguished from the character of the extreme. We have a conspicuous minimum at each solstice, but, according to Dr. Hann, the minimum of June and July is the principal, that of December and January not being nearly so noticeable. Thus, while the similarities at the solstices result in a minimum occurring at each of them, the principal minimum occurs when the Earth is most distant from the Sun, and is moving in its orbit with least velocity; the less important minimum when the converse is the case. It would seem, therefore, that the variation of the distance from the Sun, and consequent variation in orbital velocity, is not without effect.

It has to be noticed that increase in orbital velocity does not infer increase in the *variation* of local velocity. In fact, to some extent, the opposite is the case. An increase in orbital velocity reduces the *ratio* of variation in local velocity. We have taken the mean orbital velocity of the Earth to be 1,110 miles a minute, and we have seen that, at the equator, with a constant axial velocity of about $17\frac{1}{3}$ miles a minute, this results in a maximum local velocity of $1,127\frac{1}{3}$ miles a minute (being 1,110 miles *plus* $17\frac{1}{3}$ miles), and a minimum local velocity of $1,092\frac{2}{3}$ miles a minute (being 1,110 miles *minus* $17\frac{1}{3}$ miles). The ratio of the minimum to the maximum is therefore as 100 is to about $103\frac{1}{6}$. Let us suppose the orbital velocity to be, say, 100 miles a minute greater than the mean, which would make it 1,210 miles a minute. Then the maximum local velocity at the equator would be $1,227\frac{1}{3}$ miles a minute (being 1,210 miles *plus* $17\frac{1}{3}$ miles), and the minimum local velocity $1,192\frac{2}{3}$ miles (being 1,210 miles *minus* $17\frac{1}{3}$ miles). The ratio of the minimum to the maximum would, therefore, in this case, be as 100 is to about $102\frac{9}{10}$.

It would seem, however, that the *actual increase* in velocity, which results from proximity to the Sun, has an effect in increasing the magnitude of the swing; and that the *actual decrease* in velocity, which results from increased distance from the Sun, has the opposite effect. At any rate the change of orbital velocity appears to be the chief distinction between the time of perihelion and the time of aphelion. At the former, according to Dr. Hann, the amplitude of the oscillation is greater in both hemispheres than it is at the latter, the latter being, indeed, coincident with the absolute minimum in amplitude.

Apparently the particulars in which the matters relative to the Earth and the orbit, respectively, differ at the equinoxes as compared with the solstices—especially the change of position, in relation to the orbit, of the terrestrial axis and equator—result in the principal maxima occurring at the equinoxes, and the principal minima at the solstices. Similarly, the particulars in which matters differ at one solstice as compared with the other—especially the variation in the distance of the Earth from the Sun and consequent altered velocity—result in a difference in the magnitude of the minima and in the amplitude of the oscillations which occur at the respective solstices. Probably analogous differences in a minor degree, which may be similarly accounted for, occur at the respective equinoxes.

As to the seasonal variations which occur in the higher latitudes in the hours of occurrence of the phases, it is evident that, owing to the changing inclination of the Earth's axis in its relation to the orbital course, the position of the four quadrants will, in these latitudes, be considerably altered at different times of the year. Thus, the portion of the Earth within the Arctic Circle, does not, at the vernal equinox, face the direction of orbital progression during the whole rotation; while, at the autumnal equinox, it faces it all the time The converse is the case with the region within the Antarctic Circle It is clear that during the seasons when these regions respectively are sheltered from the direction of orbital progression, it can scarcely be said of them—as we have found is the case in the tropical regions—that during twelve hours in each axial rotation every geographical position is moving axially in the same direction as the Earth is moving forward in its orbit. It is equally true that at these seasons the regions in question can scarcely be said to be at any time moving axially in the opposite direction from that in which the Earth is moving onward in its orbit. The axial movement in these regions is, in fact, at the seasons specified, of such a character as gives it only a very partial resemblance, at any stage of the rotation, to a forward or backward movement in relation to the orbital progress of the Earth. Consequently, . our argument would, in its application to the Arctic Regions, at and near the vernal equinox, and in its application to the Antarctic Regions, at and near the autumnal equinox, have to be greatly modified. The exposure of the temperate regions is similarly altered, although not to the same extent. These changes have, doubtless, an important bearing on the question of the time phases. The variation in the nature of the exposure may hasten or delay, as the case may be, the time at which a barometric maximum or minimum takes place.

In view of the fact that local variations of velocity, such as we have endeavoured to describe, undoubtedly occur, it would seem reasonable to suppose that they have an important bearing on the much-vexed question of the origin of the semi-diurnal barometric oscillations. As to whether they have any bearing on the eight-hour oscillation, referred to by Dr. Hann, it is, in the present state of our knowledge of the subject, impossible to express an opinion.

There is a certain fascination about these regular daily movements of the barometer, which have for a quarter of a millennium proved so inexplicable. Theory after theory has been put forward in explanation, and has been weighed in the balance, and found wanting. While it is, of course, absolutely certain that any theory which is not consistent with known facts must be erroneous, it does not follow that a theory having a certain consistency must necessarily be correct. All the facts connected with the subject should be capable of explanation by any complete and accurate theory. On the other hand, with our imperfect information on the subject, a theory may present apparent difficulties which further investigation may dissipate. We are now groping our way towards a complete understanding of these strange phenomena, and, undoubtedly, certain progress has been made, if only in the collection of data to work upon and to reconcile with each other. Whether any one of the theories which we have endeavoured to describe will stand the test of detailed and searching enquiry, time alone will show. In any case, it must be the desire of all who are interested in the problem, that the true solution, and it only, should secure scientific acceptance. In past ages the acceptance of scientific truth has frequently been delayed through predisposition in favour of erroneous views. In these times we are coming to realize that, in the investigation of physical phenomena, all bias should be cast aside.

The truth only should be sought out. Any such predisposition cannot do more than delay the ultimate acceptance of the correct solution. In meteorological problems, as in the moral world, the old proverb stands firm—"the truth is mighty and will prevail."

THE END.

INDEX.

———o———

INDEX.

Printed and Bound by GALL & INGLIS, Newington Printing Works, Edinburgh.

Lightning Source UK Ltd.
Milton Keynes UK
UKHW010752261118
332983UK00009B/762/P

9 781334 033759